AF643301

ABHANDLUNGEN DER DEUTSCHEN AKADEMIE
DER WISSENSCHAFTEN ZU BERLIN

Jahrgang 1969

4. Internationales Symposium

Biochemie und Physiologie der Alkaloide

Halle (Saale), 25. bis 28. Juni 1969

Band a des Symposiumsberichtes
Vorabdruck wissenschaftlicher Beiträge

Herausgegeben von
KURT MOTHES, KLAUS SCHREIBER und HORST ROBERT SCHÜTTE

Redaktion
DIETER GROSS, HANS-WERNER LIEBISCH, HORST ROBERT SCHÜTTE
und URSULA STEPHAN

Mit 1 Porträt, 16 Abbildungen, 2 Tabellen und 13 Schemata

AKADEMIE-VERLAG · BERLIN
1969

Erschienen im Akademie-Verlag GmbH, 108 Berlin, Leipziger Straße 3–4

Copyright 1969 by Akademie-Verlag GmbH

Lizenznummer: 202 · 100/435/69

Offsetdruck: VEB Druckerei „Thomas Müntzer", 582 Bad Langensalza

Bestellnummer: 2001/69/II/2 · ES 18 G 1

12,–

Inhaltsverzeichnis

1819 - 1969

150 Jahre Alkaloidbegriff

Dr. Carl Friedrich Wilhelm Meissner (1792 - 1853)

Angeregt durch Sertürners Studien über das Opium bemühte sich Dr. Carl Friedrich Wilhelm Meissner, "Löwen"-Apotheker in Halle an der Saale, um die Isolierung der "wirksamen Prinzipien" verschiedener Pflanzen. Im Rahmen einer kurzen Abhandlung in Schweiggers "Journal für Chemie und Physik", 25, 379 (1819) berichtete er

II. Ueber ein neues Pflanzenalkali (*Alkaloid*).

Vom

Dr. W. Meifsner.

Darin wird ein "eigenthümlicher alkalischer Pflanzenkörper" aus Sabadillsamen beschrieben, den Meissner zur Reihe der "leicht zersetzbaren Pflanzenalkalien" wie Morphin und Strychnin zählt. Die Arbeit schließt mit einem kritischen Vergleich der Eigenschaften dieser Pflanzenstoffe und führt zu dem Vorschlag, die "Pflanzenalkalien" als Alkaloide zu bezeichnen.

Wäre es erlaubt diesem Stoff einen eigenen Namen zu geben, so würde ich vor der Hand Sabadillin vorschlagen; sollte er jedoch auch in anderen Arten von der Gattung Veratrum aufgefunden werden, so würde es wohl besser seyn ihn Veratrin zu nennen. Ueberhaupt scheint es mir auch angemessen, die bis jetzt bekannten alkalischen Pflanzenstoffe nicht mit dem Namen Alkalien, sondern Alkaloide zu belegen, da sie doch in manchen Eigenschaften von den Alkalien sehr abweichen; sie würden daher in dem Abschnitt der Pflanzenchemie vor den Pflanzensäuren ihre Stelle finden.

(From the Imperial College, London, England)

The Biosynthesis of the Aromatic Erythrina Alkaloids

by D.H.R. Barton and D.A. Widdowson

Nearly fifteen years ago (1, 2) we advanced a biosynthetic scheme for a number of alkaloids which was based on the formation of carbon-carbon and carbon-oxygen bonds by the pairing of phenolate radicals. This scheme had merit in that it not only explained the existance of many known structures but also made a number of predictions about structures that might be found in Nature.

In the last decade a number of the predicted structures have indeed been discovered but, more importantly, tracer work in our laboratories and elsewhere has provided extensive evidence for the correctness of the basic concept. A study of the reactions involved in their biosynthesis at the enzymatic level is still lacking but one anticipates with some confidence that this approach too will confirm the general correctness of the scheme.

Some typical alkaloid structures where there is now good tracer evidence for the construction of the carbon skeleton by phenolate coupling reactions are as follows. The morphine alkaloids e.g. morphine (I), sinomenine (II), galanthamine (III), haemanthamine (IV), lycorine (V), crotonosine (VI), roemerine (VII), isothebaine (VIII), epistephanine (IXa) and colchicine (IXb). In each structural formula the arrow indicates the bond or bonds which are formed by the phenolate radical coupling

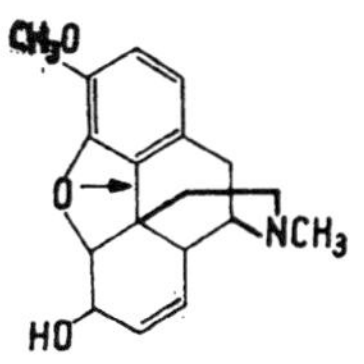

(I) *Morphine*

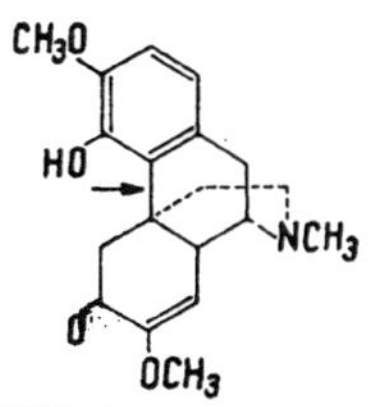

(II) *Sinomenine*

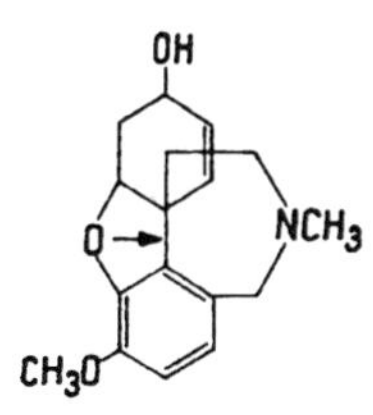

(III) *Galanthamine*

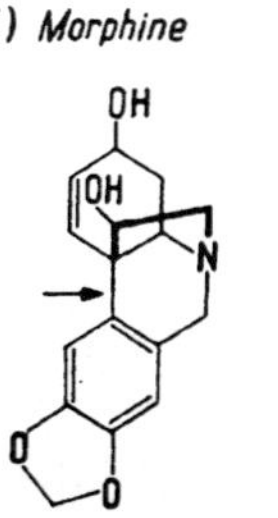

(IV) *Haemanthamine*

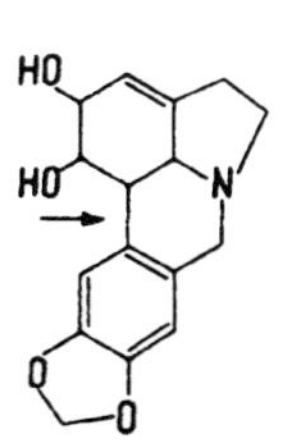

(V) *Lycorine*

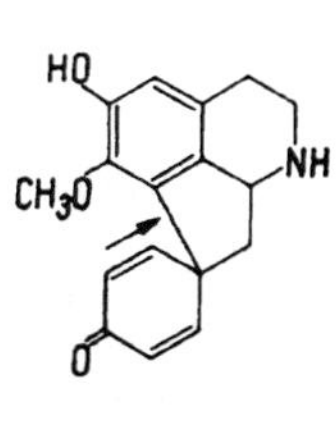

(VI) *Crotonosine*

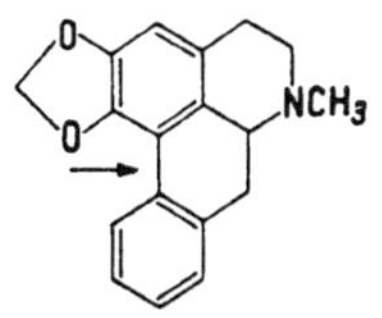

(VII) *Roemerine*

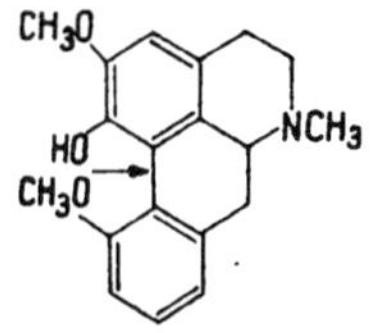

(VIII) *Isothebaine*

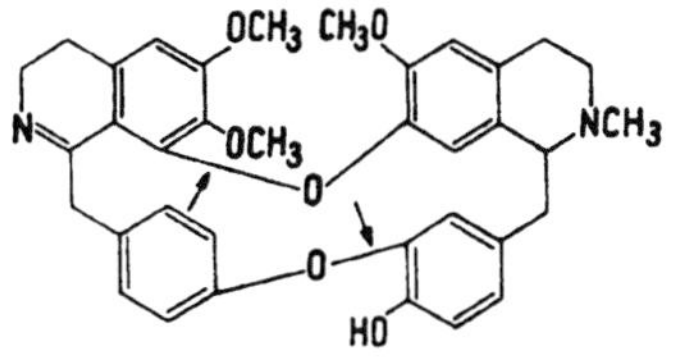

(IXa) *Epistephanine*

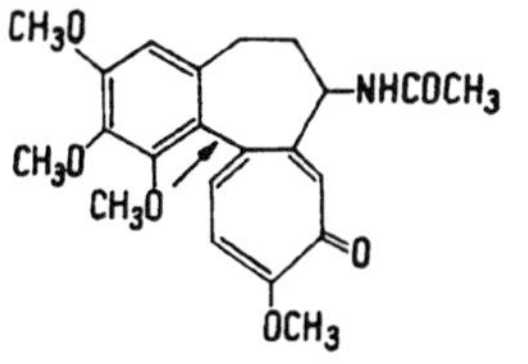

(IX b) *Colchicine*

reaction.

With the theoretical background to the role of phenol coupling reactions in biosynthesis being well understood we can proceed at once to an exposition of some of our recent work on the biosynthesis of the aromatic Erythrina alkaloids.

Since all of the more reasonable theories of Erythrina alkaloid biogenesis (1, 3) require tyrosine as the primary nitrogen precursor, we fed[2-^{14}C]-tyrosine by several methods to Erythrina crista galli and to E. rubrinervia. Acceptable incorporations into erythraline (X, R, R'=CH_2) were only achieved by the cotton wick method. Of the two species E. crista galli gave the higher incorporations (Table 1; Feedings No. 1 and 2). Consequently, cotton wick feedings to E. crista galli were used for all subsequent experiments. In all our work we have used six month old Erythrina trees.

In Erythrina alkaloids the suggested precursor, according to the theory of phenolate radical coupling (1), was the bisphenethylamine (XI). This could be envisaged (Scheme 1) to undergo intramolecular coupling to the biphenyl (XII), further oxidation of which would give the diphenoquinone (XIII). Intramolecular addition of the secondary amine function to the quinone system would then

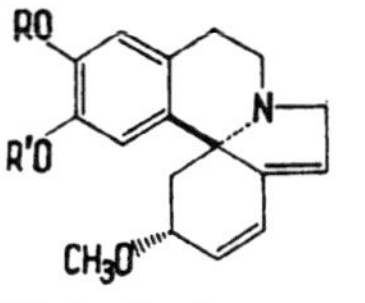

(X) *Erythraline*

R,R' = —CH_2—

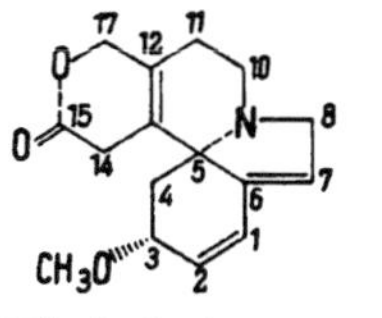

(XV) *β-Erythroidine*

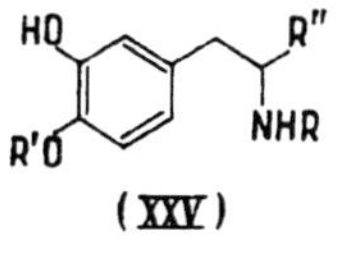

(XXV)

Table 1. Incorporation of Singly Labelled Precursors

No.	Precursor	Erythraline	Erythratine
1	[2-^{14}C]-Tyrosine	0.121	-
2	[2-^{14}C]-Tyrosine	0.031*	-
3	Bis-([2,6-^{3}H]-3-hydroxy-4--methoxyphenylethyl) amine (XI)	0.0043	0.0022
4	Bis-([2,6-^{3}H]-3-hydroxy-4--methoxyphenylethyl) amine (XI)	0.0012	0.0055
5	(±)-[5,2',6'-^{3}H]-N-Norprotosinomenine (XX, R=H)	0.24	0.020
6	(±)-[5,2',6'-^{3}H]-N-Norprotosinomenine (XX, R=H)	0.048	0.006
7	(+)-[5-^{3}H]-N-Norprotosinomenine (XXII)	0.21	0.025
8	(-)-[5-^{3}H]-N-Norprotosinomenine (XXII, Enantiomer)	0.002	0.002
9	[3,3'-^{3}H$_2$]-azapentanobiphenyl (XII)	0.46	0.21
10	(±) [1,17-^{3}H$_2$]-Erysodienone (XIV)	0.085	0.15
11	(±) [1,17-^{3}H$_2$]-Erysodienols (I and II) (XLII)	0.017	-
12	(±)[1,17-^{3}H$_2$]-Erysotinone (XXXVI)	0.31	-
13	(±)[1,17-^{3}H$_2$]-Erysotine (XXXVII)	0.18	-
14	(+)[17-^{3}H]-Erysodine (XXXVIII)	0.22	-
15	(+)[1,17-^{3}H$_2$]-Erythratinone (XL)	-	-
16	(+)[2-^{3}H]-Erythratine (XLI, β-OH)	0.041	-
17	(+)[2-^{3}H]-Erythratine (XLI, β-OH)	0.050	-
18	(+)[2-^{3}H]-Epierythratine (XLI, α-OH)	0.41	-
19	(+)[2-^{3}H]-Epierythratine (XLI, α-OH)	0.28	-

* Feeding to E. rubrinervia, all other feedings to E. crista galli

afford 'erysodienone' (XIV). Reduction, dehydration and further minor modications would then furnish the whole known series of aromatic Erythrina alkaloids.

At about the time that our programme was initiated Leete (4) reported the incorporation of [2-^{14}C]-tyrosine into β-erythroidine (XV). Degradation showed that the label was equally distributed between C-8 and C-10. Since the lactone ring of β-erythroidine should arise by cleavage of the oxygenated benzene ring of a suitable aromatic Erythrina alkaloid (5), the result implies a symmetrical intermediate derived from two tyrosine units. Such an intermediate would be the bisphenethylamine (XI).

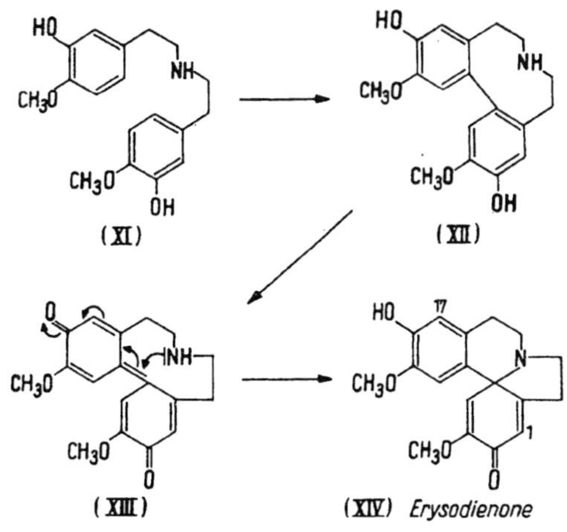

Scheme 1

Accordingly, the bisphenethylamine was synthesised as in Scheme 2. N-Benzyl-2-(3-benzyloxy--4-methoxyphenyl) ethylamine (XVII) was formed by the condensation of 2(3-benzyloxy-4-methoxy-phenyl)ethylamine (6) (XVI) with benzyaldehyde, followed by reduction of the Schiff's base with sodium borohydride or lithium aluminium hydride. Photolysis of 3-benzyloxy-4-methoxy-ω-diazoacetophen-one (7) (XVIII) in the presence of the amine (XVII) gave the amide (XIX), which was reduced with lithium aluminium hydride and hydrogenated to give the bisphenethylamine (XI). Tritiation by base

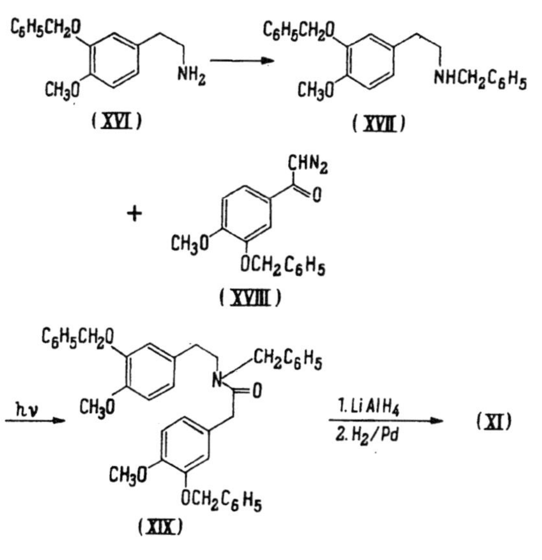

Scheme 2

catalysed exchange with tritiated water (8) gave the amine (XI) tritiated in the free o- and p-positions to the hydroxyl groups.

The tritiated amine (XI) was fed as the hydrochloride by the cotton wick method to E. crista galli plants. The incorporations (Table 1, Feeding No. 3) was however below a significant level. This could have arisen because of a lack of transport of the material to the site of synthesis, or more likely, because the compound fed was not the true precursor.

It was necessary, therefore, to consider alternative precursors. One attractive possibility was the benzyltetrahydroisoquinoline, N-norprotosinomenine (XX, R=H). p, p-Coupling of this would give the spirodienone (XXI) (Scheme 3) which on ring opening (XXI, see arrows) and reduction would generate the same biphenyl (XII) as was postulated in Scheme 1.

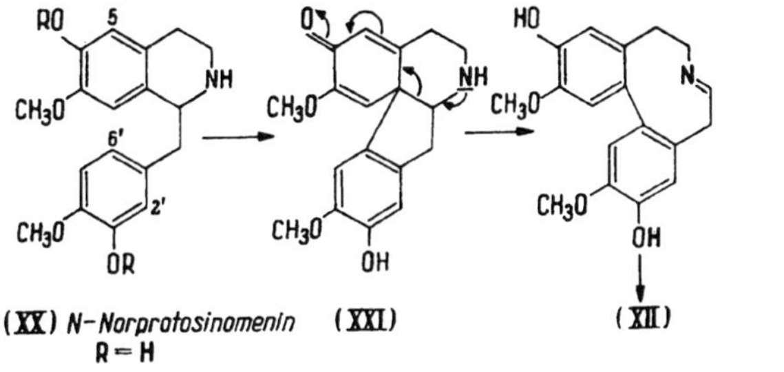

Scheme 3

0, 0-Dibenzyl-N-norprotosinomenine (XX, R=CH_2Ph) was synthesised by the Bischler-Napieralski method (7) and after debenzylation with hydrogen and palladium, was tritiated by a base catalysed exchange (8) with tritiated water at 100°. Prior deuteriation and n.m.r. analysis had confirmed (9) the exchange at positions 5, 2' and 6'.

One of the most stringent tests for a genuine enzymatic process is a marked differentiation between the percentage incorporations of optical antipodes of a given substrate. Accordingly, the synthetic 0, 0-dibenzyl N-norprotosinomenine was resolved via the two dibenzoyltartrate salts, to give the two enantiomers, $[\alpha]_D^{25}$ + 23° and -22.4°. The absolute stereochemistries of these compounds were determined by two independent methods. Firstly, the antipodes were related to laudanosine according to Scheme 4. The 0, 0-dibenzyl base was first N-methylated with formic acid/sodium formate/ formaldehyde, then debenzylated with hydrogen and palladium on charcoal and finally 0-methylated with diazomethane. By this means (-)0, 0-dibenzyl N-norprotosinomenine (XXII, R=CH_2Ph)$[\alpha]_D$-22.4°, was converted into (+)-laudanosine (XXIII)$[\alpha]_D$ + 56° (lit.[10] + 56°). The assignment was confirmed by conversion of the (+) isomer into (-)laudanosine$[\alpha]_D^{25}$ - 56°. Since the derived laudanosines were optically pure it followed that the benzylisoquinoline precursors had been fully resolved. Secondly, the (+)- and (-)-0, 0-dibenzyl N-norprotosinomenines were debenzylated with hydrogen and palladium on charcoal to give respectively (-)- and (+)-N-norprotosinomenine $\left[[\alpha]_D - 18^\circ \text{ and } + 16^\circ\right]$. The O.R.D. curve of the (-) isomer, kindly determined by Dr. P. M. Scopes, Westfield College, London, by

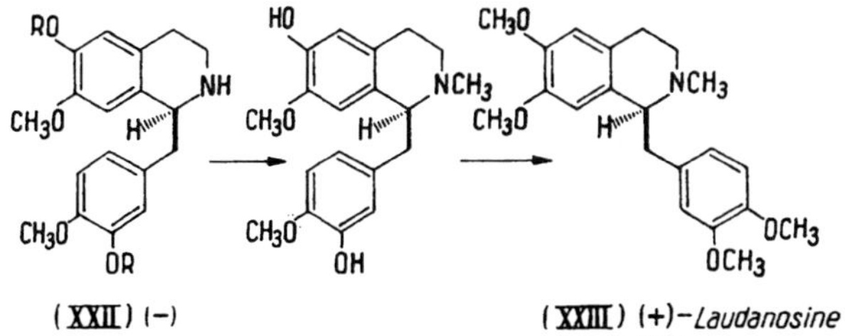

(XXII) (–) (XXIII) (+)-*Laudanosine* $[\alpha]_D^{25}$: + 56°

Scheme 4

comparison with the curves of benzyltetrahydroisoquinolines of known absolute stereochemistry (11), indicated an absolute configuration opposite to that in (XXII, R=H). This is in agreement with the above assignment of configuration to the (+)-isomer as in (XXII, R=H).

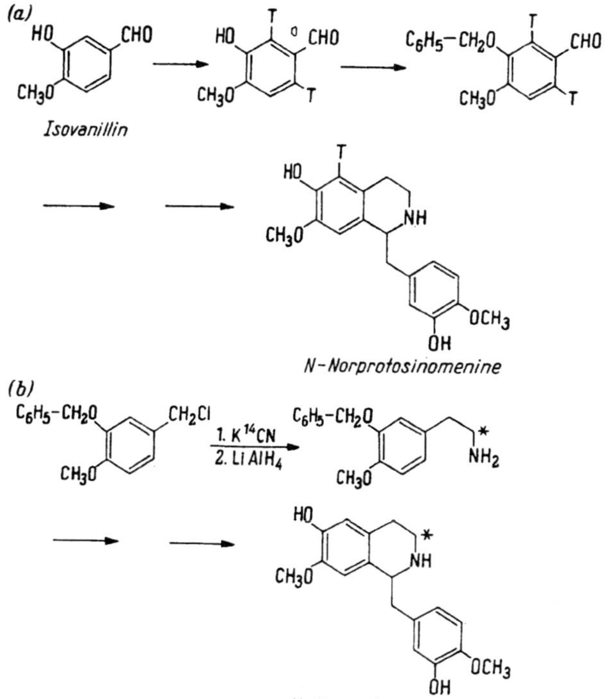

Scheme 5

The resolved bases were prepared in labelled form, starting from isovanillin (Scheme 5a). This was tritiated under basic conditions, benzylated, and converted into [5-^{3}H]-N-norprotosinomenine by the standard route (7). A variation via the nitrile route to 3-benzyloxy-4-methoxy phenylethylamine enabled 3-^{14}C -N-norprotosinomenine to be produced (Scheme 5b), by introduction of the label from K[^{14}C]CN.

The antipodes were fed to E. crista galli plants (Table 1; Feedings No. 7 and 8), and a clear distinction between the two was obtained. The (+)-isomer (XXII, R=H) (0.21% incorporation) was at least 100 times more efficient as a precursor of erythraline than the (-)-isomer, and at least 12 times more efficient as a precursor of erythratine. The reason for the lower figure for erythratine is probably a reflection of the small amount of this alkaloid in the plant and thus the difficulty of determining a percentage incorporation. The figure of 0.002% for the (-)-isomer is approaching the limit of detectability of our counter. The true figure may indeed be much lower or zero.

Final confirmation of the authenticity of N-norprotosinomenine as the precursor of the Erythrina alkaloid skeleton was secured by a double labelling experiment. [3-^{14}C, 5-^{3}H]-N-Norprotosinomenine with a ^{3}H:^{14}C ratio of 2.04 was fed (Table 2, Feeding No. 1) to E. crista galli plants. The erythraline produced (0.086% incorporation) had an ^{3}H:^{14}C ratio of 1.98, indicating a direct incorporation with no scrambling. The specificity of the labelling in this experiment will be considered later.

The proposed biosynthetic sequences (Schemes 1 and 3) both involved the biphenyl derivative (XII) and erysodienone (XIV). The synthesis of these two compounds became, therefore, of importance. Two groups of workers (12, 13) have recently described the oxidation of the bisphenethylamine (XI) by potassium ferricyanide in a two phase methylene chloride-water system to give erysodienone (XIV) in about 35% yield. We have repeated this work consistently and without difficulty. Although this oxidative cyclisation was undertaken to imitate a proposed scheme (1) of Erythrina alkaloid biosynthesis (see above), which we have now shown to be not correct, nevertheless the synthesis is of considerable importance.

The excellent yields obtained in this phenol oxidation are surprising if the biphenyl (XII) is the first intermediate. In general, the formation of nine-membered rings is a relatively inefficient process (14). In addition, if a molecular model of the diphenoquinone (XIII) is examined, it is found that the nitrogen-atom cannot easily approach the angular carbon atom as required in Scheme 1. These factors led to the suggestion (12) that the first intermediate could be the indoline (XXIV) (Scheme 6).

We have attempted to determine the correct oxidation sequence via three approaches; (i) by the synthesis and oxidation of phenethylamines of the type (XXV). (ii) by the synthesis and oxidation of the indoline (XXIV) and (iii) by the synthesis and oxidation of the biphenyl (XII) and its derivatives. Phenethylamines (XXV, R=H or CH_2Ph, R' =OMe, R"=H) were on hand. Oxidation of these under the conditions of erysodienone formation gave only polymeric material. Cyclisation of such compounds has been observed, but only when quinone formation was possible as with (15) (XXV, R=H, R'=H, R" =CO_2H) or in the Fremy's salt oxidation (16) of (XXV, R=R'=R" =H).

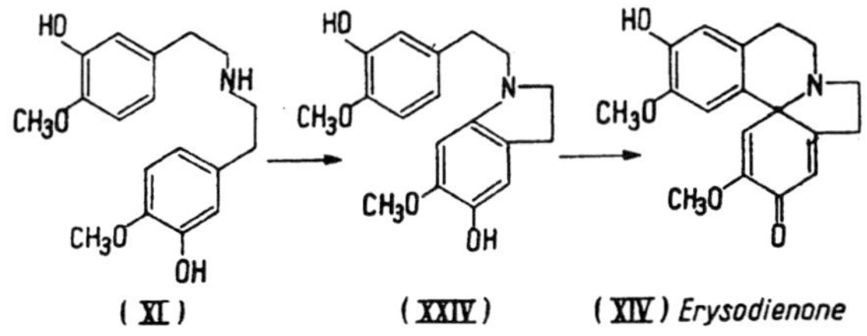

Scheme 6

The indoline (XXIV) was prepared as in Scheme 7. The bromodiazoketone (XXVI) was synthesised by standard reactions following our earlier schemes. Photolysis of this diazoketone in the presence of the appropriate phenethylamine gave the desired amide (XXVII), which on reduction with diborane (12, 13) afforded the secondary amine (XXVIII). Treatment of the latter with sodamide in liquid ammonia afforded the expected benzyne (XXIX) which at once cyclised to the indoline (17) derivative

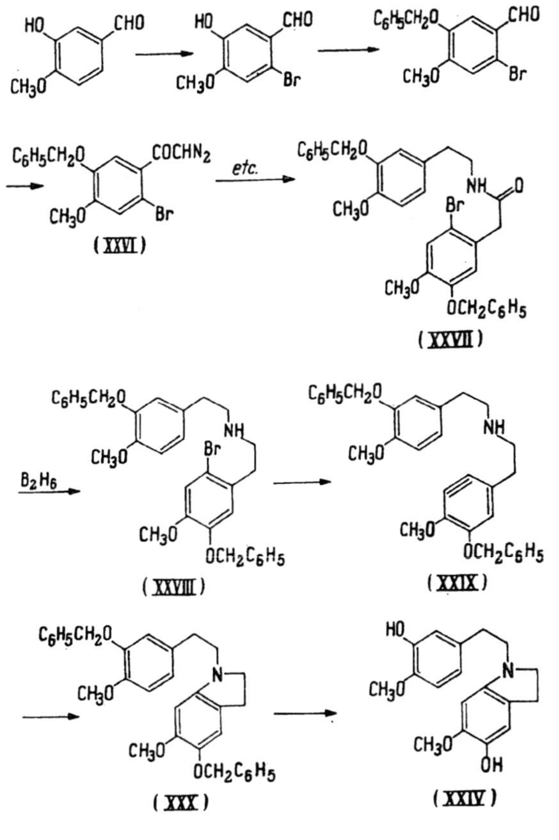

Scheme 7

(XXX). Hydrogenation in the usual way then afforded the desired indoline diphenol (XXIV). This proved to be very unstable, even in the absence of oxygen, and was usually generated from the dibenzyl derivative immediately prior to use. Oxidation of the indoline (XXIV) under the conditions of erysodienone formation again gave only polymeric tarry products and no detectable erysodienone. If this compound were to be the intermediate a yield of erysodienone of at least 35% would be necessary.

These results ruled out the formation and transformation of the indoline as the mechanism of erysodienone formation, but the original criticisms of biphenyl formation and oxidation were still valid. We therefore synthesised the biphenyl (XII) by reduction of erysodienone (XIV) (Scheme 8a).

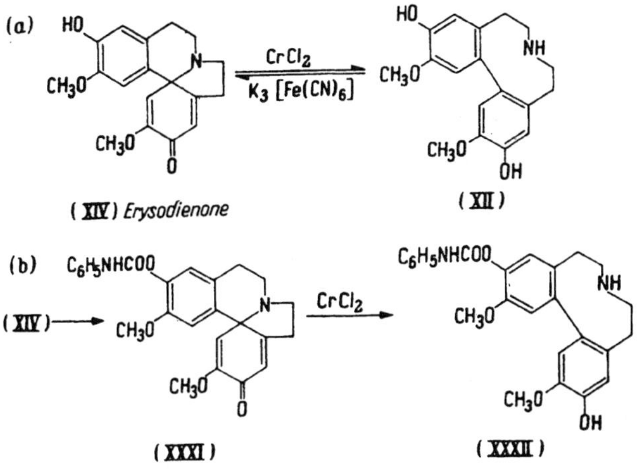

Scheme 8

This was readily accomplished with chromous chloride, or zinc and acetic acid or sodium in liquid ammonia. Oxidation with potassium ferricyanide under standard conditions gave a yield of 80% of erysodienone. This provided strong support for the intermediacy of the biphenyl (XII), but the diphenoquinone intermediate (XIII) was still in doubt. Accordingly, erysodienone (XIV) was converted to the phenylurethane (XXXI) and this reduced to the monophenylcarbamidobiphenyl (XXXII) (Scheme 8b). Attempted oxidation of this did not regenerate the dienone (XXXI), and only starting biphenyl together with some polymer was recovered. This indicated that both phenoxy groups are involved in erysodienone formation. This is only accommodated by formation of the diphenoquinone (XIII), as in Scheme 1, or by formation of a non-planar diradical (Scheme 9). We are unable to distinguish between these on the available evidence, but in view of the negative results with the phenylcarbamido-derivative, the diradical process seems an unlikely one.

Although the evidence supports Scheme 1 for the chemical process, the high yield remains unexplained. There are several factors which could exert an influence on the coupling process. These are the presence of the hetero-atom and the presence of 2 benzene rings, which could reduce the

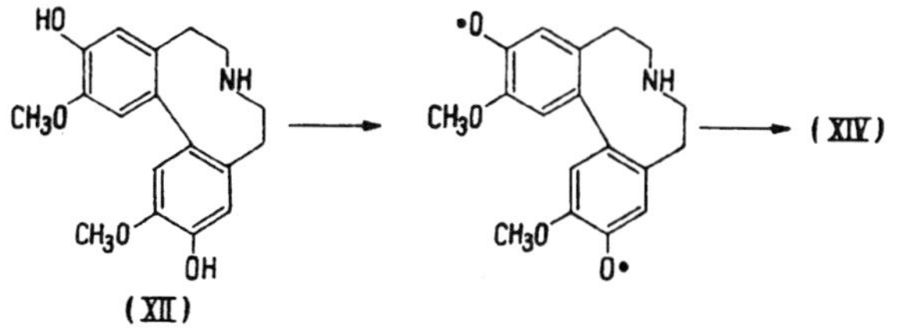

Scheme 9

conformational randomness that usually operates against the formation of a nine-membered ring. In addition, the symmetry of the system in which the phenoxy radical formation is equally facile at either end, will lead to a relatively high probability of formation of the necessary diradical (18). There is also the intriguing possibility of anion and radical exchange. If, for example, we bring a phenolate anion close in space to a phenolate radical there may be an exchange between them. In the extreme this would amount to a resonance hybrid in which the negative change and the radical were spread equally over, in the case in question, two identical rings. In either explanation a special bonding situation might be present which would facilitate the formation of a nine-membered ring by orienting the two benzene rings into a favourable conformation. Clearly all these factors require further study to assess their significance.

The biphenyl derivative (XII), now available, was tritiated by base catalysed exchange (8) and fed to E. crista galli (Table 1, Feeding No. 9). A relatively high incorporation was secured. Further confirmation of its intermediary was obtained by a double labelling experiment with the [Me-$^{14}C_2$, 5,5'-$^{3}H_2$]-biphenyl (Table 2, Feeding No. 2). The ratio of ^{3}H:^{14}C in the derived erythraline (8.9) was unchanged from that of the biphenyl (8.8). Furthermore, Zeisel determination showed that 50% of the ^{14}C label was in the O-methyl group, and acid hydrolysis gave formaldehyde (assayed as the dimedone derivative), containing 50% of the ^{14}C label. The compound was therefore incorporated without scrambling and the methylene dioxy ring was generated by cyclisation of the methoxyl group (19).

Table 2. Multiple Label Feedings

No.		Ratio ($^{3}H/^{14}C$)	Erythraline (%)	Ratio ($^{3}H/^{14}C$)	^{14}C-Label CH$_2$	^{14}C-Label MeO
1	[3-^{14}C, 5-^{3}H]-N--Norprotosinomenine (XXII)	2.04	0.086	1.98	-	-
2	[Me-^{14}C, -3,3'-^{3}H]-azapentanobiphenyl (XII)	8.8	0.39	8.9	50.5%	50.5
3	[Me-^{14}C-1,17$^{3}H_2$]--Erysodienone (XIV)	10.3	0.53	10.4	50.1	48.9

Erysodienone (XIV) was tritiated by base catalysed exchange (8). Equivalent deuteriation and n.m.r. analysis showed that exchange occurs at positions 1 and 17. The tritiated material was incorporated into erythraline (X) with adequate efficiency (0.085%). Better proof of its intermediary was secured by a double labelled feeding [Me-$^{14}C_2$, 1,17-$^{3}H_2$]-erysodienone, (Table 2; feeding No. 3), in which the label ratio (10.3) was unchanged in erythraline (10.4), and all of the ^{14}C label was shown to be equally distributed between the methoxy and methylene dioxy-groups.

All of the above biosynthetic evidence points to the correctness of the new N-norprotosinomenine postulate, but so far the steric course of the reactions have been ignored. Consider now the stereochemistry of the (+)-N-norprotosinomenine precursor in relation to the derived erysodienone (Scheme 10), which must have the same absolute stereochemistry (20) at C-5 as erythraline (X; R, R =CH_2),

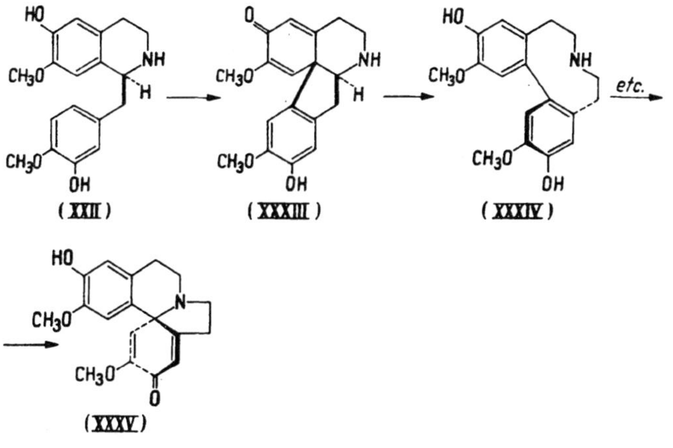

Scheme 10

i.e. the stereostructure must be as (XXXV). (+)-N-Norprotosinomenine (XXII) would give the spirodienone (XXXIII) on oxidative coupling. Ring opening and reduction of this would produce the biphenyl of the indicated stereostructure (XXXIV). If the asymmetry of the biphenyl is retained, then either an inversion has to occur in proceeding to erysodienone, or a symmetrical intermediate is involved, which must then proceed to the dienone by selection of the opposite asymmetry to that of the original biphenyl.

We have sought more direct evidence for the steric nature of the intermediates by feeding the unsymmetrically labelled [3-^{14}C, 5-^{3}H]-norprotosinomenine. If the assymetry were retained, and an inversion took place, the ^{3}H label would be entirely at C-1 or C-17 of the dienone (there are two possible modes of inversion), and the ^{14}C label would be entirely at C-8 or C-10. If a symmetrical intermediate were involved, then equal labelling at these positions would result.

These experiments are still in hand and definitive results have not been obtained at the time of writing this manuscript.

Finally, it is pertinent to consider the further transformations of erysodienone to the known aromatic alkaloids. This can be expressed by the general Scheme 11. Feeding results (Table 1, Feeding Nos. 11-19) conform to the Scheme, where there are no stereochemical ambiguities. In erysotine (XXXVII) and erythratine (XLI), the hydroxyl groups can be α- or β. In the case of erysotine, which is a synthetic material (13), it was not possible to resolve the epimeric pair if indeed both epimers were present so that the configuration of the true intermediate has not been determined, but erythratine (XLI, β-OH) and epi-erythratine (XLI, α-OH) are readily separable. The incorporations of this epimeric pair, (Table 1, Feedings Nos. 16-19) fed in parallel, indicate that epi-erythratine is probably the true precursor of erythraline, and erythratine merely a byproduct. It is possible that the elimination of water from erythratine and its epimer is a 'chemical' process, rather than a strictly enzymatically controlled one, since the biosynthetic behaviour is exactly paralleled by the chemical behaviour. Methanesulphonyl chloride/pyridine converts erythratine to erythraline in only 13% yield, but epierythratine (which has the hydroxy group axial) is converted quantitatively.

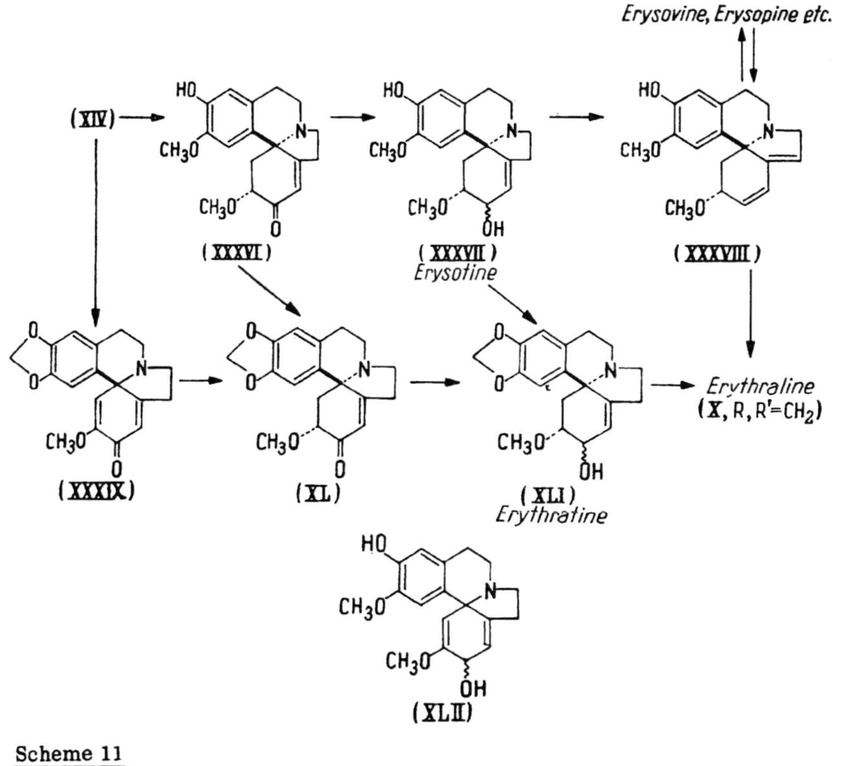

Scheme 11

The epimeric mixture of erysodienols (13) (XLII) again as yet unseparable, was incorporated into erythraline at very low efficiency. Until this mixture is separated and fed individually the intermediacy of the erysodienols between erysodienone and erysotinone must remain in doubt.

References

(1) Barton, D.H.R., and T. Cohen, Festschrift A. Stoll, Birkhäuser A.-G., Basel, 1957, p. 117;

(2) Barton, D.H.R., Chem. in Brit. 3, 330 (1967);
Taylor, W.I., and A.R. Battersby, Oxidative Coupling of Phenols, Marcel Dekker Inc., New York, 1968;

(3) Witkop, B., and S. Goodwin, Experientia (Basel) 8, 337 (1952);
Boekelheide, V.; J. Weinstock, M.F. Grundon, G.L. Sauvage and E.J. Agnello, J. Amer. chem. Soc. 75, 2550 (1953);
Wenkert, E., Chem. and Ind. 1953, 1088;
Boekelheide, V., and V. Prelog, Progr. org. Chem. 3, 259 (1955);
Boekelheide, V., Rec. Chem. Prog. 16, 227 (1955);
Robinson, R., "Structural Relations of Natural Products", Clarendon Press, Oxford 1955, p. 84;
Prelog, V., Angew. Chem. 69, 33 (1957);

(4) Leete, E., and A. Ahmad, J. Amer. chem. Soc. 88, 4722 (1966);

(5) van der Linden, A.C., and G.J.E. Thijsse, Advances in Enzymol. 27, 469 (1965);

(6) Robinson, R., and S. Sugasawa, J. chem. Soc. (London) 1931, 3163;

(7) Barton, D.H.R., A.J. Kirby and G.W. Kirby, J. chem. Soc. (London) (C) 1968, 929;

(8) Kirby, G.W., and L. Ogunkoya, J. chem. Soc. (London) (C) 1965, 6914

(9) Barton, D.H.R., R. James, G.W. Kirby, D.W. Turner and D.A. Widdowson, Chem. Commun. 1966, 295

(10) Corrodi, H., and E. Hardegger, Helv. chim. Acta 39, 889 (1956)

(11) Battersby, A.R., I.R.C. Bick, W. Klyne, J.P. Jennings, P.M. Scopes and M.J. Vernengo, J. chem. Soc. (London) 1965, 2239

(12) Gervay, J.E., F. McCapra, T. Money, G.M. Sharma and A.I. Scott, Chem. Commun. 1966, 142

(13) Mondon, A., and M. Ehrhardt, Tetrahedron Letters (London) 1966, 2557

(14) Finley, K.T., Chem. Rev. 64, 573 (1964)

(15) Cromartie, R.I.T., and J. Harley-Mason, J. chem. Soc. (London) 1952, 2525

(16) Teuber, H.-J., and O. Glosauer, Chem. Ber. 98, 2648 (1965)

(17) Kametani, T., and K. Ogasawara, J. chem. Soc. (London) (C) 1967, 2208

(18) Da Rooge, M.A., and L.R. Mahoney, J. org. Chemistry 32, 1 (1967)

(19) Barton, D.H.R., G.W. Kirby, J.B. Taylor and G.M. Thomas, J. chem. Soc. (London) 1963, 4545

(20) Boekelheide, V., and G.R. Wenzinger, J. org. Chemistry 29, 1307 (1964)

(From the Department of Medicinal Chemistry and Pharmacognosy,
Purdue University, Lafayette, Indiana, U.S.A.)

The Biosynthesis of Ergot Alkaloids

By H. G. Floss

Introduction

During the 4 years since the third symposium in 1965, a considerable number of publications have dealt with ergot alkaloids from various viewpoints. Several new alkaloids have been isolated and their structures have been determined. A new pharmacological activity of an ergot alkaloid has been uncovered with the finding that agroclavine is a powerful inhibitor of lactation. Several laboratories have published work on the mechanism of the formation of ergot alkaloids from tryptophan and mevalonic acid, on biogenetic relationships among these alkaloids, and on aspects of the physiology and regulation of their formation. Yet, despite the considerable amount of information on ergot alkaloid biosynthesis that has been accumulated (cf. ref. 1-5 for recent reviews) one cannot overlook the fact that our knowledge of this biosynthetic pathway is still fragmentary. In fact, it seems as if some of the recent results have even complicated matters rather than clarified them.

Formation of the Tetracyclic Ergoline Ring System

The biogenetic precursors of the ergolines, tryptophan, mevalonic acid, and the methyl group of methionine, have been recognized for quite some time (6). Mevalonic acid is apparently incorporated via γ,γ-dimethylallyl pyrophosphate, which has been shown to give rise to radioactive alkaloids when fed to the cultures in ^{14}C-labeled form (7). The incorporation of the latter compound, which is somewhat surprising in view of the well known impermeability of cell membranes to phosphate esters, is not due to dephosphorylation and utilization of the alcohol, since γ,γ-dimethylallyl alcohol-1-T is not specifically incorporated into ergot alkaloids (8). It has become quite clear now that γ,γ-dimethylallyl pyrophosphate reacts with the 4-position of the indole moiety of tryptophan to give rise to 4-(γ,γ-dimethylallyl)-tryptophan (9). Plieninger's group (7) has presented additional evidence for the intact incorporation of this compound into ergot alkaloids, and more recently Róbbers (10), as well as Agurell and Lindgren (11) have demonstrated its formation by the ergot fungus under conditions which inhibit alkaloid synthesis. Furthermore, Plieninger et al. (7) have shown that 4-(γ,γ-dimethylallyl)-tryptamine is incorporated into ergot alkaloids. However, whether this compound is a true intermediate in the pathway is still questionable (7, 10). While the mechanism of the formation of ring C is still unclear, Fehr, Acklin and Arigoni (12, 13) have shown that deoxy-chanoclavine-I and deoxy-nor-chanoclavine-I are not converted into tetracyclic ergolines. Consequently, one of the methyl groups of the dimethylallyl residue must be hydroxylated before ring C is closed.

The role of chanoclavines in alkaloid biosynthesis, a matter of controversy for some years, has now been clarified. Three groups have reported that chanoclavine-I is a very efficient precursor of the tetracyclic ergolines (14-16). Isochanoclavine-I which has 5 R, 10 R configuration like almost all the tetracyclic ergolines and chanoclavine-I, but contrary to the latter carries the hydroxymethyl group cis to the carbon substituent at C-9 (17, 18), and chanoclavine-II are not converted into tetracyclic ergot alkaloids (15, 19, 20). Chanoclavine-I, as well as the other two chanoclavine isomers, carries the label from mevalonic acid-2-^{14}C in the methyl group (15) and its cyclization is accompanied by a cis-trans isomerization of the double bond giving rise to agroclavine and elymoclavine which are labeled at C-17 (15, 19, 20). It follows from experiments with mevalonic acid-4 R, 4 T that the original dimethylallyl residue should be labeled in the trans methyl group (20, 21). Therefore, the initial hydroxylation must take place at the cis methyl group, which by another cis-trans isomerization then becomes the trans hydroxymethyl group of chanoclavine-I. These relationships are summarized in figure 1. Some studies have been made on the mechanism of the cyclization/iso-

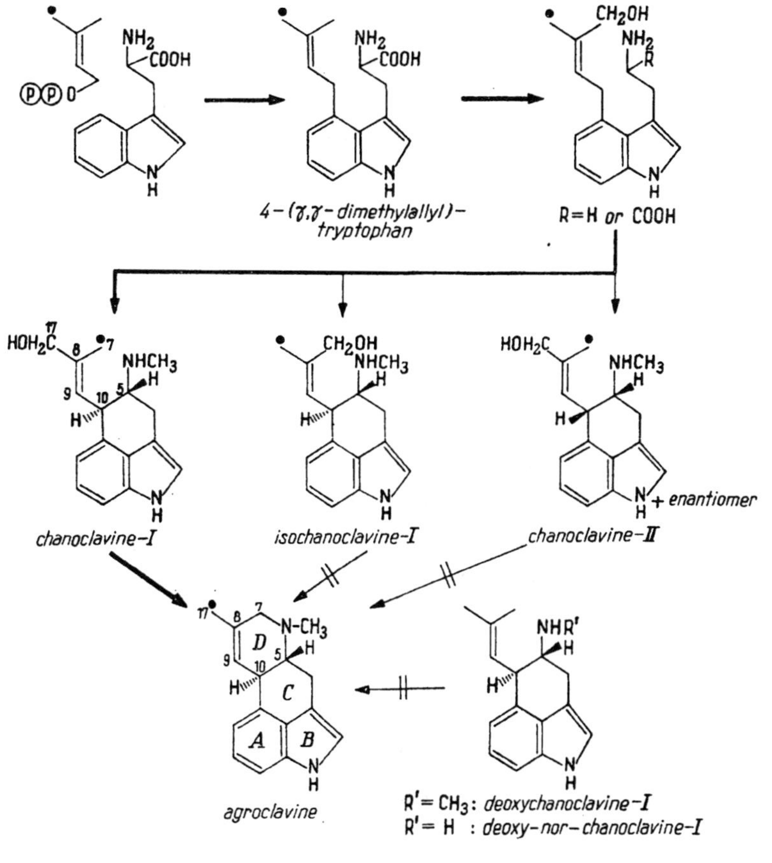

Figure 1. Formation of the ergoline ring system (⟶ main pathway)

merization of chanoclavine-I. Tritium at C-10 is completely retained during this process, but chanoclavine-I-7-^{14}C-9-T was cyclized with a 30 % decrease in the T/^{14}C ratio (20). The results of feeding experiments with mevalonic acid-3', 4-D_5, in which both chanoclavine-I and tetracyclic ergolines were analyzed for their deuterium content, are interpreted as indicating (a) an intermolecular transfer of the hydrogen at C-9 and (b) loss of one of the methylene hydrogens from the hydroxymethyl group of chanoclavine-I (22).

Biogenetic Interrelationships of the Alkaloids

Clavine alkaloid interconversions by the ergot fungus have been studied most extensively by Agurell and Ramstad (23). The main pathway of biosynthesis involves hydroxylation of agroclavine to elymoclavine (23), which in turn is converted into lysergic acid derivatives (24, 25) (figure 2). In addition, a great number of secondary transformations of these alkaloids take place, such as double bond reduction, isomerization, dehydration, and hydroxylations. One general type of transformation is the 8-hydroxylation of agroclavine and elymoclavine, giving rise to the diastereomeric setoclavines and penniclavines, respectively. This reaction, unlike the methyl group hydroxylation of agroclavine, can be brought about by a great variety of organisms (26), and recently it has been shown by the groups of Ramstad and Taylor to be catalyzed by peroxidase (27, 28). The initial enzymatic attack occurs at C-10 and the resulting 10-hydroxy derivative then undergoes an allyl rearrangement (28):

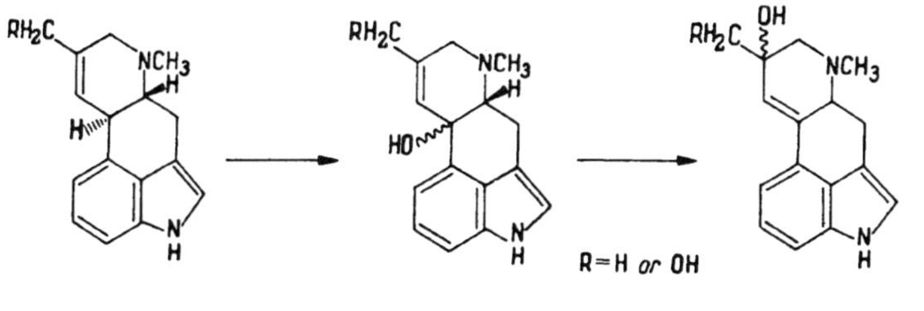

On the other hand, the hydroxylation of agroclavine to elymoclavine is not catalyzed by peroxidase (29). The hydroxyl oxygen of elymoclavine, as well as that of chanoclavine-I, must originate from molecular oxygen (30), but attempts to carry out this hydroxylation in a cell-free system have so far been unsuccessful (31, 29).

The detailed pathway of the conversion of elymoclavine to lysergic acid derivatives is not clear yet, but it has been established that neither the lysergols, nor penniclavine are intermediates (32, 33). Agurell's studies on the biosynthesis of lysergic acid α-hydroxyethyl amide in Paspalum ergot suggest that elymoclavine is dehydrogenated to 6-methyl-$\Delta_{8,9}$-ergolene-8-carboxylic acid, a natural ergot constituent (34), which then undergoes a double bond shift (35). Of particular importance was the finding that lysergamide and isolysergamide are not converted into their corresponding α-hydroxyethyl amides (36). According to further experiments, ethylamine is not incorporated into the α-hydroxyethyl moiety (36), and pyruvate and alanine are the most specific precursors of this two-carbon portion (1, 37, 38). Although somewhat obscured by the rapid metabolic turnover of alanine, there is evidence that the amino acid nitrogen preferentially provides the amide nitrogen of the α-hydroxyethyl amide (38). Majer et al. (39) found a low incorporation of alanine, but no incorporation of L-alaninol into the three-carbon side-chain of ergometrine. They also observed incorporation of L-alanine and pyruvate into peptide alkaloids, but the isotope distribution indicated extensive randomization of the label and no preferential incorporation of radioactivity into the α-hydroxyalanine moiety of ergotamine. All these data are in agreement with a suggestion by Agurell (1) that lysergylalanine is a common precursor of lysergic acid α-hydroxyethyl amide, ergometrine, and the ergotamine type peptide alkaloids (figure 2). Experimental evaluation of this hypothesis (40) has so far shown incorporation (1.77 %) of lysergyl-L-alanine-(alanine-2-^{14}C) into

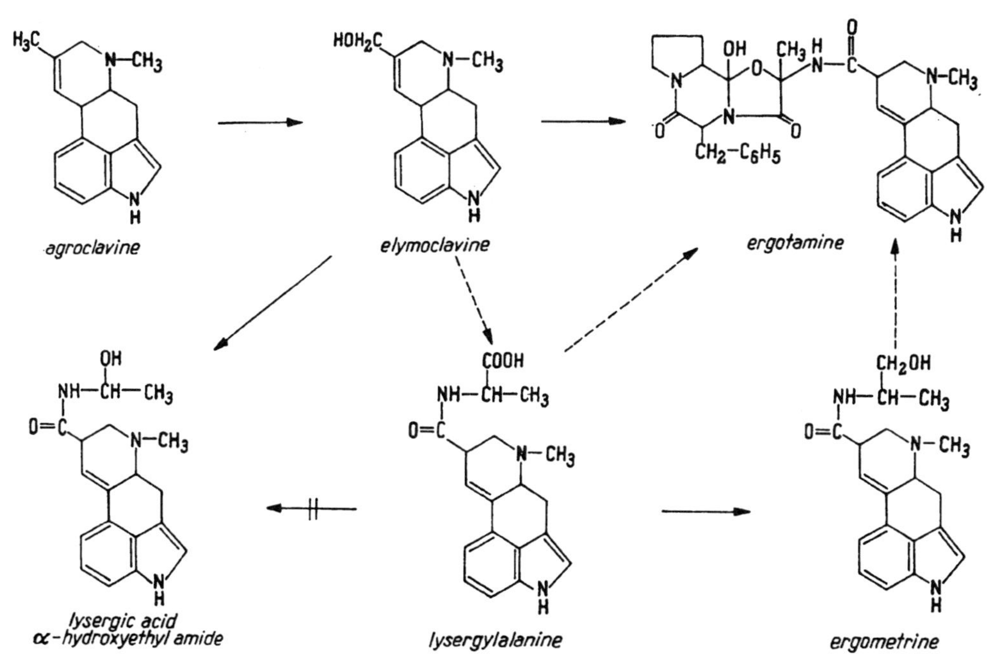

Figure 2. Biogenetic relationships among ergot alkaloids
(——→demonstrated, - - - - -→ postulated)

ergometrine (97.5 % of the radioactivity in C-2 + C-3 of alaninol), but not into the acetaldehyde moiety of the α-hydroxyethyl amide (< 0.09 %). Since both alkaloids were isolated from the same cultures and since these cultures were free of alkaloid at the begin of the experiment, permeability problems or lack of synthesis of a particular alkaloid cannot account for the negative result. It appears, therefore, that lysergylalanine is only a precursor of ergometrine but not of lysergic acid α-hydroxyethyl amide. Non-incorporation of lysergylalanine into the latter has also been observed by Agurell (41). Its incorporation into ergotamine has not yet been examined. Likewise, the assumption of Voigt and Bornschein (42) that ergometrine is an intermediate in the formation of peptide type alkaloids, which resulted from their nonisotopic precursor feeding experiments, has not yet been examined directly by using labeled ergometrine.

Physiology of Alkaloid Formation

Of the many studies on this subject only a few can be mentioned here, and the selection is obviously quite arbitrary. Teuscher (43) has demonstrated a correlation between the ability of Claviceps strains to produce alkaloids and their ability to actively take up L-tryptophan from the medium. Only alkaloidproducing strains were shown to actively transport the amino acid into the cells. Studying a Claviceps purpurea strain which produces ergotamine in submerged culture, Amici et al. found that this strain is heterokaryotic, i.e., its cells contain several different nuclei. It seggregates into three apparently monokaryotic strains which all produce very little or no alkaloid. From two of the new strains they were able to synthesize a heterokaryon which resembled the original strain in many respects and did produce a considerable amount of alkaloid. Based on their work they "suggest that alkaloid production is related to heterokaryosis of the producing strain" (45).

Based on work by U.Mothes it has been suggested that alkaloid synthesis in Claviceps is an inducible process and that tryptophan acts as the inducer (46). This would explain the findings that tryptophan stimulates alkaloid formation when added at the begin of the culture period, but not when added after alkaloid production has started, that a variety of tryptophan analogues stimulates alkaloid production even though they are not incorporated into the alkaloids, and that mycelium grown in the presence of tryptophan or its analogues retains the ability to produce more alkaloid than the controls even after replacement into new medium not containing the effectors. In agreement with this idea, it has been shown that protein is synthesized at a considerable rate during the alkaloid production phase (47, 48, 49), although no net increase in the amount of protein is observed (49). BuLock and co-workers (48) have noticed fluctuations in the rate of alkaloid synthesis which they also explain on the basis of the above assumption. In addition, they found that alkaloid synthesis ceases within 24 hours after blocking protein synthesis. Further work in our laboratory (50) showed that the level of free tryptophan in the mycelium temporarily increases 2-3 fold during the transition from growth to alkaloid production phase (figure 3).

As pointed out by Lingens, the stimulation of alkaloid production by tryptophan analogues (46) also suggests that tryptophan biosynthesis in Claviceps may be controlled less tightly than usual, since most of these analogues are effective feedback regulators of tryptophan biosynthesis in other organisms (51). His studies (52), which showed lack of feedback inhibition of anthranilate synthetase from Claviceps paspali by tryptophan, have added weight to this assumption. We (53) have examined the levels of tryptophan synthetase in the mycelium of Claviceps strain SD 58 in the presence and absence of exogenous tryptophan (figure 4). Both during the growth phase and the alkaloid production phase there is a 4 - 6 fold repression of the enzyme by added tryptophan. During the transition from growth to alkaloid production phase, the enzyme level increases 20 - 25 fold in normal as well as in tryptophan-supplemented cultures. The inerease in the latter cultures is not due to exhaustion of exogenous tryptophan. These data are difficult to interpret. They are clearly not in accord with a model of simple end-product repression, but as one explanation one could assume that the formation of a separate tryptophan synthetase is induced together with the alkaloid-synthesizing enzymes.

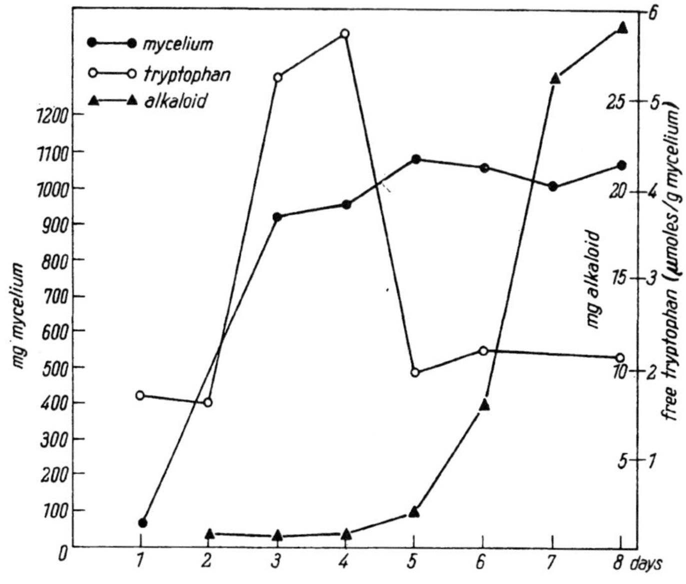

Figure 3. Growth, alkaloid production, and levels of mycelial free tryptophan in cultures of Claviceps spec., strain SD 58

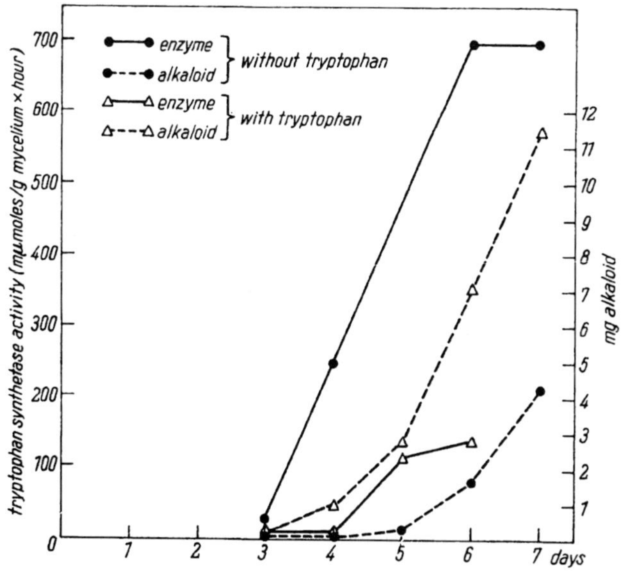

Figure 4. Alkaloid and tryptophan synthetase levels in cultures of Claviceps strain SD 58 in normal and tryptophan-supplemented ($2 \cdot 10^{-3}$ M) medium

In trying to summarize these observations and to put the pieces together one may speculate about the events preceding the onset of alkaloid formation in the following way: Cessation of growth reduces the rate of protein synthesis and leads to an overproduction of amino acids which in the case of tryptophan is not immediately counteracted by feedback regulation. Strains possessing tryptophan permease will build up a higher concentration of free tryptophan in the cells, which then induces the synthesis of the enzymes for alkaloid formation and possibly also a separate set of enzymes of the tryptophan pathway. As alkaloid synthesis commences the concentration of free tryptophan decreases again. The supply of mevalonate is apparently never limiting nor is this compound able to induce alkaloid formation, because we were never able to increase alkaloid synthesis by adding mevalonate.

Related Work

Recently, Holzapfel (54) isolated the toxic principle from strains of Penicillium cyclopium and showed that the compound, named cyclopiazonic acid, has structure I (figure 5). In continuation of these elegant studies, Holzapfel and Wilkins (55) isolated β-cyclopiazonic acid (figure 5, II), a precursor of

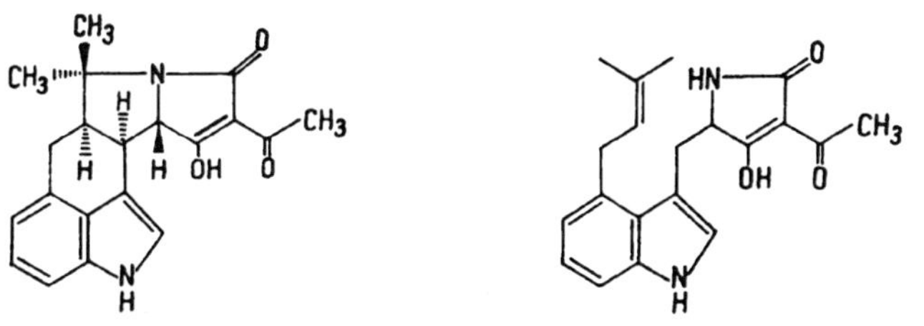

(I) *cyclopiazonic acid (configuration shown is relative)* (II) *β-cyclopiazonic acid*

Figure 5. Metabolites from Penicillium cyclopium

cyclopiazonic acid, from the same organism. The biogenetic similarity to the ergolines, which is suggested by the structures of I and II, was confirmed by feeding labeled mevalonate, tryptophan, and acetate, and by degradation of the resulting I. The 4 additional carbon atoms attached to the amino acid side chain represent two acetate units. Apparently, the formation of cyclopiazonic acid and of ergolines represent two ways in which the 4-(γ, γ-dimethylallyl)-tryptophan system can undergo cyclization.

Acknowledgments

Work from the author's laboratory was generously supported by the National Institutes of Health (Grant No. AM 11662) and by Eli Lilly and Company.

References

(1) Agurell, S., Acta Pharm. Suecica **3**, 71 (1966)

(2) Gröger, D., Fortchr. Chem. Forsch. **6**, 159 (1966)

(3) Voigt, R., Pharmazie 23, 289, 353, 419 (1968)

(4) Gröger, D., and H.G. Floss, Herba Hungarica, in print

(5) Ramstad, E., Lloydia, in print

(6) cf. Weygand, F., and H.G. Floss, Angew. Chem. 75, 783 (1963)

(7) Plieninger, H., H. Immel, and A. Völkl, Liebig's Ann. Chem. 706, 223 (1967)

(8) Günther, H.F., and H.G. Floss, unpublished results

(9) Plieninger, H., R. Fischer, and V. Liede, Liebig's Ann. Chem. 672, 223 (1964)

(10) Robbers, J.E., and H.G. Floss, Arch. Biochem. Biophysics 126, 967 (1968)

(11) Agurell, S., and J.-E. Lindgren, Tetrahedron Letters [London] 1968, 5127

(12) Arigoni, D., "Some Aspects of Mevalonoid Biosynthesis". Symp. on Organic Chemical Approaches to Biosynthesis, London 1965

(13) Fehr, T., Ph. D. Dissertation, ETH Zürich 1967

(14) Gröger, D., D. Erge, and H.G. Floss, Z. Naturforsch. 21b, 827 (1966)

(15) Fehr, T., W. Acklin, and D. Arigoni, Chem. Commun. 1966, 801

(16) Voigt, R., M. Bornschein, and G. Rabitzsch, Pharmazie 22, 326 (1967)

(17) Stauffacher, D., and H. Tscherter, Helv. chim. Acta 47, 2186 (1964)

(18) Acklin, W., T. Fehr, and D. Arigoni, Chem. Commun. 1966, 799

(19) Floss, H.G., U. Hornemann, N. Schilling, D. Gröger, and D. Erge, Chem. Commun. 1967, 105

(20) Floss, H.G., U. Hornemann, N. Schilling, K. Kelley, D. Gröger, and D. Erge, J. Amer. chem. Soc. 90, 6500 (1968)

(21) Floss, H.G., Chem. Commun. 1967, 804

(22) Floss, H.G., unpublished

(23) Agurell, S., and E. Ramstad, Arch. Biochem. Biophysics 98, 457 (1962)

(24) Mothes, K., K. Winkler, D. Gröger, H.G. Floss, U. Mothes, and F. Weygand, Tetrahedron Letters [London] 1962, 933

(25) Gröger, D., H.R. Schütte, and K. Stolle, Z. Naturforsch. 18b, 1850 (1963)

(26) Beliveau, J., and E. Ramstad, Lloydia 29, 234 (1966); and references therein

(27) Taylor, E.H., and H.R. Shough, Lloydia 30, 197 (1967)

(28) Chan Lin, W.-N., E. Ramstad, and E.H. Taylor, Lloydia 30, 202 (1967)

(29) Jindra, A., E. Ramstad, and H.G. Floss, Lloydia 31, 190 (1968)

(30) Floss, H.G., H. Günther, D. Gröger, and D. Erge, J. Pharmac. Sci. 56, 1675 (1967)

(31) Tyler, V.E., D. Erge, and D. Gröger, Planta med. [Stuttgart] 13, 315 (1965)

(32) Agurell, S., and M. Johansson, Acta chem. scand. 18, 2285 (1964)

(33) Floss, H.G., H. Günther, D. Gröger, and D. Erge, Z. Naturforsch. 21b, 128 (1966)

(34) Kobel, H., E. Schreier, and J. Rutschmann, Helv. chim. Acta 47, 1052 (1964)

(35) Agurell, S., Acta Pharm. Suecica 3, 65 (1966)

(36) Agurell, S., Acta Pharm. Suecica 3, 33 (1966)

(37) Castagnoli, N., and A. Tonolo, IX. Int. Congr. Microbiology, Moscow 1966, Symposia p. 31

(38) Gröger, D., D. Erge, and H.G. Floss, Z. Naturforsch. 23b, 177 (1968)

(39) Majer, J., J. Kybal, and T. Komersova, Folia microbiol. [Praha] 12, 489 (1967)

(40) Basmadjian, G., H.G. Floss, D. Gröger, and D. Erge, unpublished

(41) Agurell, S., private communication

(42) Voigt, R., and M. Bornschein, Pharmazie 19, 772 (1964)

(43) Teuscher, E., Flora (Jena) 155, 80 (1964)

(44) Amici, A.M., A. Minghetti, T. Scotti, C. Spalla, and L. Tognoli, Appl. Microbiol. 15, 597 (1967)

(45) Amici, A.M., T. Scotti, C. Spalla, and L. Tognoli, Appl. Microbiol. 15, 611 (1967)

(46) Floss, H.G., and U. Mothes, Arch. Mikrobiol. 48, 213 (1964)

(47) Rothe, U., and W. Fritsche, Arch. Mikrobiol. 58, 77 (1967)

(48) Barr, J.G., J.D. BuLock, and D. Shepherd, Lloydia, in print

(49) Kaplan, N., U. Hornemann, and H.G. Floss, unpublished; cf. H.G. Floss, Ber. dtsch. bot. Ges. 80, 705 (1967)

(50) Basmadjian, A., and H.G. Floss, unpublished

(51) Lingens, F., Hoppe-Seyler's Z. physiol. Chem. 331, 56 (1963)

(52) Lingens, F., W. Goebel, and H. Uessler, Europ. J. Biochem. 2, 442 (1967)

(53) Jindra, A., K.M. Kelley, and H.G. Floss, unpublished

(54) Holzapfel, C.W., Tetrahedron [London] 24, 2101 (1968)

(55) Holzapfel, C.W., and D.C. Wilkins, 5th International Symposium on the Chemistry of Natural Products, London July 1968, abstract C 65

(From the Department of Botany, University College London, England)

Unusual Amino Acids from Plants

By L. Fowden

General considerations

Just twenty years ago we started working with the peanut, a plant that was to launch my laboratory on its continuing quest for novel amino acids in plants. Like many other workers in natural product research, we had adopted paper chromatography, then in its early days, as a routine tool for surveying large numbers of plant extracts for unusual constituents. Today, other more sophisticated chromatographic techniques have replaced paper chromatography in many fields of research, but the original method is still preferred by most workers when amino acids are under investigation.

The peanut gave γ-methyleneglutamic acid and γ-methyleneglutamine (1) as two of the early examples of non-protein amino acids (2) from plants. In 1950, such chemical characterizations were based upon evidence derived (a) from classical chemical methods for recognizing functional groups, (b) from i.r. spectroscopy, and (c) by synthesis. Now, in the amino acid field as in others, the potential of n.m.r. and mass spectroscopy has been harnessed with a concomitant reduction both in the time and in the quantity of material necessary for unequivocable structural predictions.

During these two past decades the number of rigidly-characterized, non-protein amino acids has increased from less than ten to more than 200, and so another significant group has been added to the list of plant secondary products. The members of this group exhibit several novel structural features not encountered among the 20 protein amino acids. For example, in the general structure $RCH(NH_2)CO_2H$, it is not uncommon to find that R may contain an ethylenic linkage [as in γ-methylene- and γ-ethylidene-glutamic acids (1, 3)], while three acetylenic amino acids have been isolated recently from seed of Euphoria longan (4) (see below). Likewise, the cyclopropyl ring system forms a part of R in seven natural amino acids obtained mainly from members of the family Sapindaceae and allied species (5), but for structural variety the heterocyclic amino and imino acids are unequalled. Imino acids based on the azetidine, pyrrolidine and piperidine rings number more than twenty, and include several unsaturated compounds, whilst a group of β-substituted alanines include compounds with heterogeneous substituents based on the following general types: pyridine [in mimosine (6) and nicotianine (7)], pyrimidine [in lathyrine (8) and willardiine (9)], pyrone [in stizolobic and stizolobinic acids (10)], pyrazole (11) and thiazole (12).

On the occasions when a particular non-protein amino acid represents a major proportion of the soluble-nitrogen fraction of a plant, it is quite common to find that the plant also contains the corresponding γ-glutamyl peptide, usually as a minor constituent; in rare instances, especially in storage organs, the peptide may be predominant, e.g. γ-glutamyl-S-methyl-cysteine in seed of some Astragalus species (13), or γ-glutamyl-β-(methylene-cyclopropyl)alanine in akee (Blighia sapida) fruits (14).

Only a few of the non-protein amino acis show a wide distribution throughout the plant kingdom: γ-aminobutyric acid is such a compound being present in almost all species the author has examined. α-Aminoadipic acid, pipecolic acid and δ-acetylornithine are other examples of amino acids occurring in members of many plant families. More often particular substances seem to be restricted to a very few species, frequently ones that are botanically closely-allied. For example, β-pyrazol-1-ylalanine is a constituent of the seeds of many cucurbit species (15), but its presence is unknown outside the family Cucurbitaceae. Similar phytochemical relationships can be confined to smaller groups of plants, i.e. between species forming a single genus as with lathyrine in Lathyrus (16). Azetidine-2-carboxylic acid for long had been regarded as such a taxonomic marker in respect of the Liliaceae, but Miss M.-L. Sung of my laboratory has just isolated the imino acid from a legume seedling. Its presence in two quite unrelated families, one monocotyledonous and the other dicotyledonous, may suggest that two distinct biogenetic pathways producing azetidine-2-carboxylic acid have been evolved by the different plant types. A somewhat parallel situation exists with γ-methyleneglutamic acid and related compounds; they are produced by a few legumes, by a small group of related genera within the Liliaceae, and by a few other species scattered rather sporadically through the plant kingdom.

Seeds or fleshy storage organs of plants frequently represent the principal sites at which accumulations of particular non-protein amino acids occur, and quite commonly the same substance either may be undetectable or present only in trace amounts in rapidly-growing green tissues. Such behaviour suggests that the amino acid may act as a reserve or storage compound during periods of dormancy or metabolic inertness: in this context, it may be relevant that canavanine, homoarginine, lathyrine and α, γ-diaminobutyric acid, which form the predominant free amino acids of seeds of various Lathyrus and Vicia species, each have nitrogen contents considerably higher than those associated with protein molecules. The considerable number of N-acetylated amino acis, e.g. δ-N-acetylornithine (17), β-N-acetyldiaminopropionic acid (18) and N-acetyldjenkolic acid (19), may have a similar role, for acetylation could represent a mechanism for temporarily withdrawing a compound from active metabolism. However, some substances such as homoserine in peas or γ-methyleneglutamic acid and γ-methyleneglutamine in peanuts are formed only after germination of the seeds, and in these instances the compounds are considered to play an important role as agents in nitrogen transport within the young plants. Perhaps it is not surprising that, within a group of some 200 structurally heterogeneous compounds, no common function has emerged; indeed, as with the alkaloids, many of these non-protein amino acids may be expected to remain without known function.

The main problems of biogenesis are concerned with processes elaborating the requisite carbon skeletons. Often the ultimate step in a biosynthetic pathway would seem to involve a transamination reaction between a compound like glutamic acid and the α-keto acid analogue of the given amino acid. For the many compounds that are either homologues or simple substituted dervatives of normal protein constituents, biogenesis may involve pathways analogous to those concerned in intermediary amino acid metabolism, e.g. γ-hydroxy-γ-methylglutamic acid could be formed by transamination from γ-hydroxy-γ-methyl-α-ketoglutaric acid, itself formed by a condensation reaction between either two molecules of pyruvate or between one of pyruvate and one of oxaloacetate in a manner reminiscent of the condensing enzyme system yielding citrate prior to α-ketoglutarate. However, the same C_6 branched chain results following a 'meta cleavage' of protocatechuate by certain strains of Pseudomonas (20). It is conceivable that two such totally unrelated pathways might provide the biogenetic duality consistent with the occurrence of γ-methyleneglutamic acid in diverse families like the Liliaceae and the Leguminosae (see above).

Labelling experiments indicate that a number of the non-protein amino acids have low turnover rates that do not facilitate exact biosynthetic studies. However, biogenetic hypotheses sometimes can be constructed by examining the 'comparative anatomy' of the structures of amino acids isolated from the same or closely-allied species of plant. The aim is to recognize compounds which, although apparently metabolic end-products in one species, can be envisaged as logical biosynthetic precursors of related compounds in an allied species (see example below for species within the Sapindaceae). This approach has a considerable similarity to the mutant technique employed so successfully by Beadle and Tatum (21) to elucidate amino acid biosynthetic pathways in Neurospora.

An increasing interest is developing in the possibility that particular non-protein amino acids may act as metabolic analogues, i.e. as alternative substrates for or inhibitors of enzymes concerned in intermediary metabolic pathways leading to the synthesis or degradation of some of the protein amino acids. To be effective in this way, the molecules of the two substances must be virtually 'isosteric', and also exhibit similar ionization behaviour. Any difference of molecular size should err towards the analogue being the smaller, and not the larger molecule. This is clearly reflected in the comparative behaviour of azetidine-2-carboxylic acid and pipecolic acid as substrates for the prolyl-sRNA synthetase purified from mung bean seeds: the former small molecule is significantly activated by this enzyme, but pipecolic acid (the larger homologue of the normal substrate) is not used as a substrate (22). The development of ideas on analogue behaviour eventually may establish some of the non-protein amino acids as regulatory substances governing aspects of amino acid metabolism in the plants producing them.

The amino acids of the Sapindaceae and Aesculus

Some taxonomists place the genus Aesculus within the Sapindaceae while others establish it as the monogeneric family Hippocastanaceae. Evidence of amino acid composition now presented, in suggesting that some Aesculus species have a closer kinship to members of certain genera within the Sapindaceae than to other Aesculus species, favours the former alternative.

Sources and structures

The first unusual amino acid to be isolated from a member of the Sapindaceae was β-(methylenecyclopropyl)alanine (I, hypoglycin A) (23). This cyclopropyl amino acid was obtained as the toxic principle responsible for the hypoglycaemic activity associated with unripe akee fruits; hypoglycin A is present in the fleshy aril as well as the seed. The homologue, α-(methylenecyclopropyl)glycine (II), was obtained later from the seed of another sapindaceous plant, Litchi chinensis, and shown to possess hypoglycaemic activity against mice (24). Quite recently, a related group of five amino acids were characterized as constituents of A. californica (California buckeye) seed (25). The C_7 branched carbon skeleton encountered in hypoglycin A evidently forms the structural basis of this new group of compounds: homoisoleucine (III), 2-amino-4-methylhex-4-enoic acid (IV), its γ-glutamyl derivative (VI), 2-amino-6-hydroxy-4-methylhex-4-enoic acid (V) and β-(methylenecyclopropyl)-β-methylalanine (VII). Compound IV is present in high concentration ($>$5 g/kg fresh seed) and represents about half of the seeds' soluble-nitrogen pool, but compounds III, VI and VII form very minor constituents (see Fig. 1).

Another species, A. parviflora, was the source of two further cyclopropyl compounds. Cis-α-(carboxycyclopropyl)glycine (IX), together with glutamic acid, formed the principal acidic amino acids present in the seed, but exo(cis)-3,4-methanoproline (X) was undoubtedly the major component of the free amino acid pool. The latter compound produces a more brilliant yellow colour than proline when chromatograms are treated with ninhydrin and, like azetidine-2-carboxylic acid, it bleaches isatin. Chromatographic evidence indicates that trans-α-(carboxycyclopropyl)glycine (XI) forms a minor constituent in A. parviflora, although isolation and characterization of this diastereoisomer of IX has been effected from Blighia sapida seed (5).

Our most recent work has concerned Euphoria longan, the sapindaceous species showing the closest resemblance to Litchi. Longan seed has yielded the first examples of acetylenic amino acids having the following structures: 2-amino-4-methylhex-5-ynoic acid (VIIIa), 2-amino-4-hydroxymethylhex-5-ynoic acid (VIIIb) and 2-amino-4-hydroxyhept-6-ynoic acid. The main evidence for structural assignments has come from hydrogenation studies yielding either the corresponding saturated (Adam's Pt catalyst) or olefinic (Lindlar catalyst) amino acids, and by n.m.r. and mass spectroscopic examination of the original acetylenic and derived olefinic compounds (4). In compounds VIIIa and VIIIb the branched C_7 structure is once again evident.

Biosynthetic considerations

The small variation between the structures of the several different C_7 amino acids described may be rationalized with their involvement in a common biogenetic pathway. One possible scheme is outlined in Fig. 1, whose purpose is to suggest a sequence in which the different compounds may be derived rather than a definite scheme of one-step metabolic conversions between compounds. Assuming that homoisoleucine (III) is the primary C_7 product, then a chain-elongating mechanism commencing from isoleucine would seem to be the most likely biosynthetic mechanism, since there are a number of analogies for this in conventional amino acid metabolism, e.g. valine to leucine, or methionine to homomethionine (as a pre-requisite for sinigrin production in Brassica nigra) (26). Other possible routes to homoisoleucine might involve leucine or tiglic acid as precursor molecules (see Fig. 1). Desaturation of homoisoleucine could yield the related olefin IV, or the acetylenic amino acid VIIIa, and their subsequent hydroxylation might give compounds V and VIIIb, respectively.

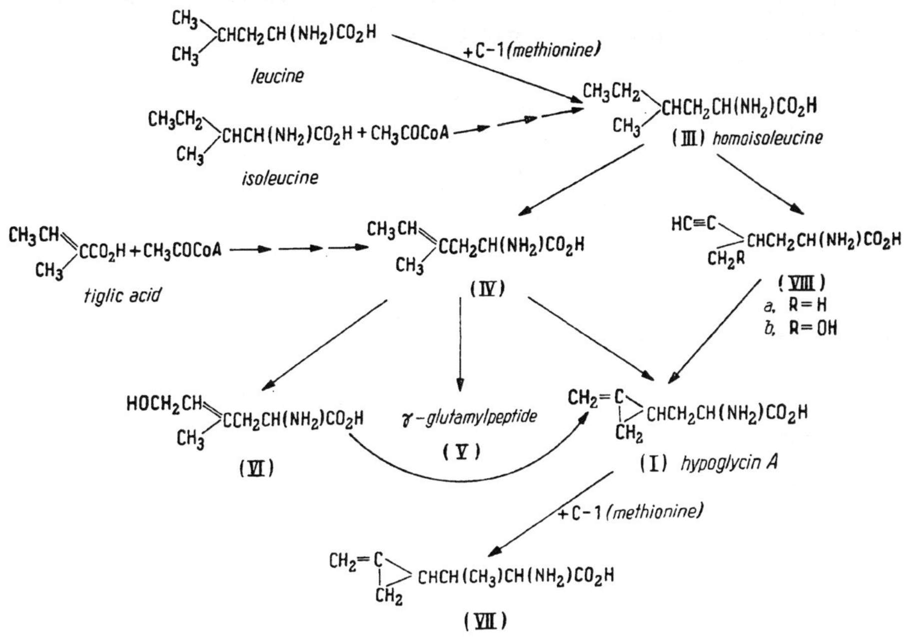

Figure 1. Possible biosynthetic pathways in Aesculus californica and related species

Hydroxylation of the C_7 skeleton probably would favour cyclization to give the cyclopropyl ring, i.e. hypoglycin A (I), but some form of additional activation reaction may be necessary. A methyl transfer from methionine to hypoglycin A would form a logical final step in the biosynthesis of β-(methylenecyclopropyl)-β-methylalanine (VII). If this scheme correctly represents the essential features of amino acid biosynthesis in A. californica, then a re-examination of this species might be expected to reveal hypoglycin A as a minor constituent.

Experimental data pertinent to Fig. 1 is very limited. ^{14}C-Isoleucine is incorporated into hypoglycin A when supplied to slices of akee aril or seed, but comparable experiments with seed slices of A. californica have not shown an introduction of label into 2-amino-4-methylhex-4-enoic acid. These experiments were performed with seed air-freighted to Britain, but they now need to be repeated and extended with akee in Jamaica and buckeye in California. Dr. R. Suhadolnik also has worked on hypoglycin A biosynthesis, finding that ^{14}C-label from pyruvate or acetate is incorporated into the amino acid, but only into the C-1 and C-2 atoms: this accords with an involvement of acetate in a chain-elongating process of the type proposed.

α-(Methylenecyclopropyl)glycine (II) and cis- (IX) and trans-α-(carboxycyclopropyl)glycine (XI) differ only in the oxidation level of the single C substituent attached to the cyclopropyl ring. In II, the asymmetry of a ring C atom is lost, and it would seem feasible for this compound to act as an intermediate in a series of reactions leading to epimerization of cis- and trans-α-(carboxycyclopropyl)glycines (see Fig. 2). (Epimerization through a non-asymmetrical intermediate is an established feature of the interconversion of UDP-glucose and UDP-galactose.) In the tentative scheme cis- and trans-forms of α-(hydroxymethylcyclopropyl)glycine are implicated as other intermediates and their status as natural products warrants investigation.

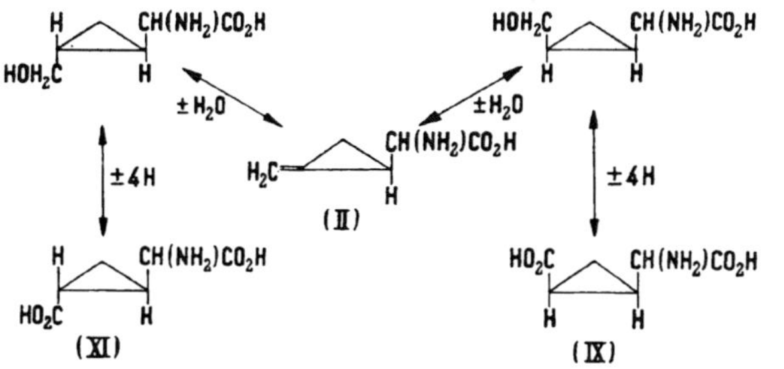

Figure 2. Possible scheme for epimerization of cis- and trans- forms of α-(carboxycyclopropyl)glycine

The recognition of cis-3,4-methanoproline (X) as a plant product could have widespread biogenetic implications. Its structural relationship to cis-α-(carboxycyclopropyl)glycine (IX) is identical to that existing between proline and glutamic acid, and it seems highly probable that dehydrogenase-type enzymes effect ring closure between the carboxyl and glycine moiety attached to the cyclopropyl ring in IX to give first an unsaturated Δ-pyrrolidine precursor of X, and finally X itself. Methanoproline also may act as a key intermediate in the interconversion of a group of plant imino acids as illustrated in Fig. 3. Isomerization reactions could link it with baikiain (XII) and 4-methyleneproline (XIII),

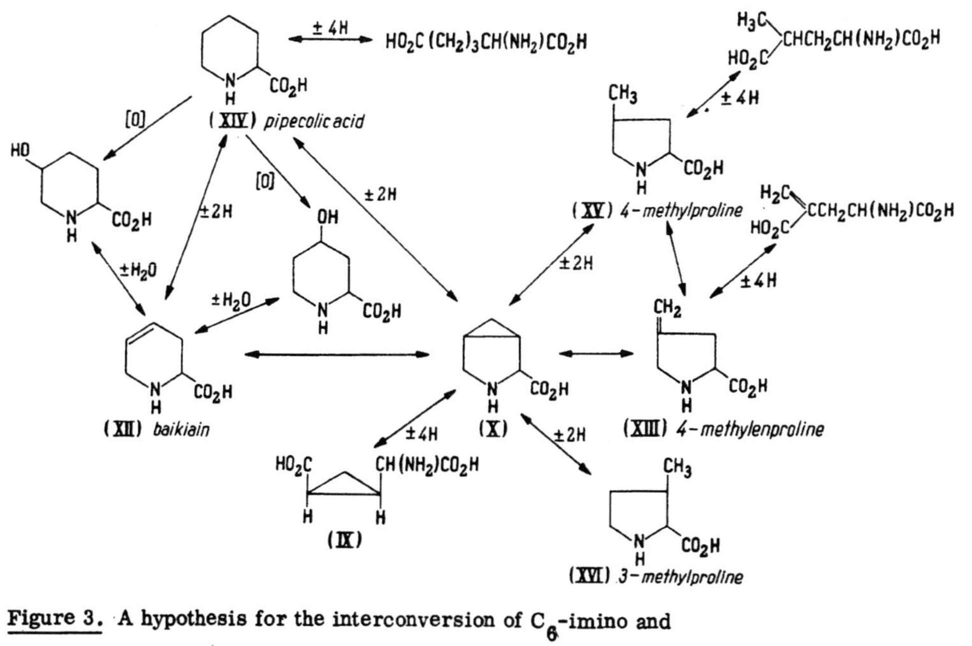

Figure 3. A hypothesis for the interconversion of C_6-imino and dicarboxylic amino acids

whilst dehydrogenase catalyzed reactions could interconvert pipecolic acid (XIV), 4-methylproline (XV) and 3-methylproline (XVI) via methanoproline. Fig. 3 also outlines other possible metabolic relationships between certain imino acids and their corresponding open-chain dicarboxylic amino acids.

Taxonomic considerations

The species forming the Aesculus genus are grouped into four sub-genera: Hippocastanum, Calothyrsus, Pavia and Macrothyrsus. The table summarizes data concerning the amino acid complement of seeds of various species of Aesculus, but only in relation to the compounds described above.

Table

Distribution of some amino acids in species of Aesculus

Compound no.*)	III	IV	V	VI	VII	IX	X	XI
I. Hippocastanum								
A. hippocastanum	0	0	0	0	0	0	0	0
A. carnea	0	0	0	0	0	0	0	0
A. turbinata	0	0	0	0	0	0	0	0
II. Calothyrsus								
A. californica	W	S	W	W	W	0	0	0
A. indica	W?	0	W	0	0	0	0	0
III. Pavia								
A. glabra	0	0	0	0	0	M	0	W
A. pavia	0	0	0	0	0	T	0	0
A. octandra	0	0	0	0	0	W	0	W
A. hybrida	0	0	0	0	0	T	0	0
A. arguta	0	0	0	0	0	M	0	M
IV. Macrothyrsus								
A. parviflora	0	0	0	0	0	M	S	T

*) see text for compound identification.

The amount of each compound present in seeds is expressed on a relative scale:

S, strong spot; M, medium; W, weak; T, trace; and 0, not detected.

Clearly, species from the four sub-genera show distinct differences in amino acid composition. For example, A. parviflora (Macrothyrsus) alone produces methanoproline (X), although all members of the sub-genus Pavia contain the suggested precursor, cis-α-(carboxycyclopropyl)glycine (IX), and often also trans-α-(carboxycyclopropyl)glycine (XI). The species forming Hippocastanum generally lack all these newly described compounds, whilst the C_7 complex of amino acids (compounds III-VII) are confined to the sub-genus Calothyrsus (as represented by A. californica and A. indica). These distribution studies then lend support to the accepted division of the Aesculus genus.

The relationship between the Aesculus species forming the Calothyrsus sub-genus and the genera Blighia and Euphoria of the Sapindaceae have been stressed at the biogenetic level in Fig. 1. Species assigned to Pavia and Macrothyrsus also share with Blighia the ability to elaborate trans-α-(carboxycyclopropyl)glycine.

Analogue behaviour

Shive and coworkers (27) have suggested that certain olefinic C_6 and C_7 amino acids may behave as phenylalanine analogues. 2-Amino-4-methylhex-4-enoic acid (IV), in which the two methyl groups bear a cis configuration about the double bond, is such an example, since the β-carbon atom is coplanar with those beyond which then resemble an incomplete phenyl ring. Now my colleague, Dr. J.W. Anderson, has demonstrated that this unsaturated amino acid can act as a substrate for purified preparations of the phenylalanyl-sRNA synthetase from both A. californica and A. hippocastanum. The V_{max} determined for the analogue is almost identical with that for phenylalanine itself. Another of my colleagues, Mr. H. Tristram, has found that 2-amino-4-methylhex-4-enoic acid forms a potent and specific inhibitor of ^{14}C-phenylalanine uptake by bacterial cells, seemingly by competing for sites on the permease system. Although less predictable, hypoglycin A and 2-amino-4-methylhex-5-ynoic acid also limited the uptake of ^{14}C - phenylalanine by Escherichia coli.

3,4-Methanoproline (X) exerts strong growth inhibition on cultures of E. coli; the effect is reversed specifically by addition of proline to the growth medium. Methanoproline also acts as a strong specific inhibitor of the proline permease system of this organism. A slight activation of methanoproline was effected by a purified prolyl-sRNA synthetase preparation from Phaseolus aureus seed.

Cis-α-(carboxycyclopropyl)glycine markedly inhibits the growth of mung bean seedlings, but the inhibition could not be alleviated by either glutamic acid or proline, so the site of anti-metabolite action of this compound awaits investigation.

References

(1) Done, J., and L. Fowden, Biochem. J. 51, 451 (1952)

(2) Fowden, L., Endeavour 21, 35 (1962)

(3) Fowden, L., Biochem. J. 98, 57 (1966)

(4) Sung, M.-L., L. Fowden, D.S. Millington, and R.C. Sheppard, unpublished

(5) Fowden, L., A. Smith, D.S. Millington, and R.C. Sheppard, Phytochemistry (in press)

(6) Renz, J., Hoppe-Seyler's Z. physiol. Chem. 244, 153 (1936)

(7) Noguchi, M., H. Sakuma, and E. Tamaki, Phytochemistry 7, 1861 (1968)

(8) Bell, E.A., and R.G. Foster, Nature (London) 194, 91 (1962)

(9) Gmelin, R., Hoppe-Seyler's Z. physiol. Chem. 316, 164 (1959)

(10) Hattori, S., and A. Komamine, Nature (London) 183, 1116 (1959)

(11) Noe, F.F., and L. Fowden, Biochem. J. 77, 543 (1960)

(12) Waisvisz, J.M., M.G. van der Hoeven, and B. te Nijenhuis, J. Amer. chem. Soc. 79, 4524 (1957)

(13) Dunnill, P.M., and L. Fowden, Phytochemistry 6, 1659 (1967)

(14) Hassall, C.H., and D.I. John, J. chem. Soc. (London) 1960, 4112

(15) Dunnill, P.M., and L. Fowden, Phytochemistry 4, 933 (1965)

(16) Bell, E.A., Biochem. J. 83, 225 (1962)

(17) Reuter, G., Flora (Jena) 145, 326 (1957)

(18) Seneviratne, A.S., and L. Fowden, Phytochemistry 7, 1039 (1968)

(19) Gmelin, R., A. Kjaer, and P.O. Larsen, Phytochemistry 1, 233 (1962)

(20) Dagley, S., P.J. Geary, and J.M. Wood, Biochem. J. 109, 559 (1968)

(21) Beadle, G.W., and E.L. Tatum, Proc. nat. Acad. Sci. USA 27, 499 (1941)

(22) Peterson, P.J., and L. Fowden, Biochem. J. 97, 112 (1965)

(23) Ellington, E.V., C.H. Hassall, J.R. Plimmer, and C.E. Seaforth, J. chem. Soc. (London) 1959, 80

(24) Gray, D.O., and L. Fowden, Biochem. J. 82, 385 (1962)

(25) Fowden, L., and A. Smith, Phytochemistry 7, 809 (1968)

(26) Chisholm, M.D., and L.R. Wetter, Canad. J. Biochem. 42, 1033 (1964)

(27) Edelson, J., C.G.Skinner, J.M. Ravel, and W. Shive, J. Amer. chem. Soc. 81, 5150 (1959)

(From the Department of Chemistry, University of Minnesota,
Minneapolis, Minnesota, U.S.A.)

The biosynthetic origin of the nitrogen in the heterocyclic rings of alkaloids

By E. Leete

A large amount of information has been obtained on the origin of alkaloids by carrying out feeding experiments with hypothetical organic precursors labelled with ^{14}C or ^{3}H. Until recently the origin of the nitrogen present in alkaloids has received little direct attention. The only isotope of nitrogen which is of practical use in tracer experiments is ^{15}N (the half-life of the radioactive isotope ^{13}N is only 10 minutes). Good methods are now available for the assay of ^{15}N, either in a mass spectrometer (1, 2, 3) or by spectrocopic methods (4, 5, 6). It seems preferable at this time to carry out mass spectral determinations for excess nitrogen on the nitrogen gas obtained from the alkaloid or its degradation products since high precision for the precentage excess ^{15}N was not obtained from the mass spectra of the organic nitrogen compounds (3). Since the natural abundance of ^{15}N is 0.37 %, high dilutions of the administered precursor cannot be tolerated.

Tracer work involving the use of ^{15}N labelled precursors in the study of alkaloid biosynthesis is recorded in the table. The most significant results have been obtained by the use of doubly labelled precursors (usually with ^{14}C), and the specific incorporation of the ^{14}C and ^{15}N, when available, is recorded in the table.

Table

Species Precursor	Alkaloid	Specific % Incorporation ^{14}C	 ^{15}N	Reference
Adhatoda vasica				
anthranilic acid-^{15}N	peganine		5.4	7
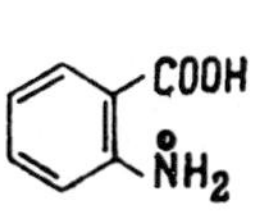	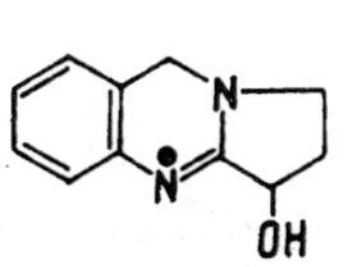			
glutamine-amide-^{15}N	peganine		3.88	7

Species Precursor	Alkaloid	Specific % Incorporation ^{14}C	 ^{15}N	Reference
aspartic acid-^{15}N	peganine	-	0.84	7
anthranilamide-amide-^{15}N	peganine	-	0.19	7
Atropa belladcnna				
ornithine-2-^{14}C,α-^{15}N	hyoscyamine	3.4	1.1	8
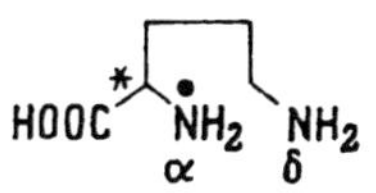	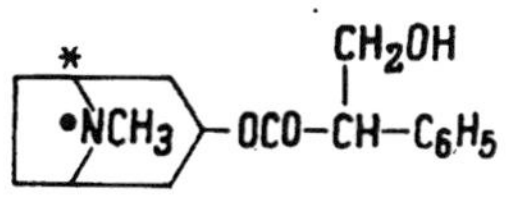			
	hyoscine	3.2	0.9	8
	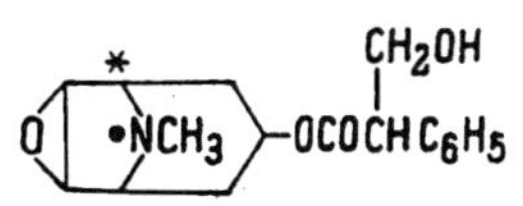			
ornithine-2-^{14}C,δ-^{15}N	hyoscyamine	5.7	1.8	8
	hyoscine	6.6	1.9	8
Cinchona succirubra				
tryptophan-2-^{14}C,1-^{15}N	quinine	0.97	0.97	9
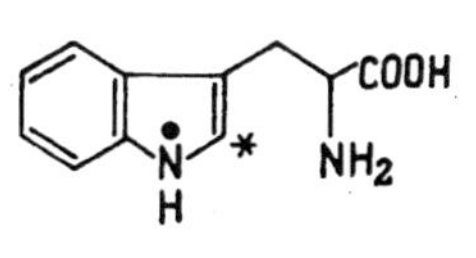	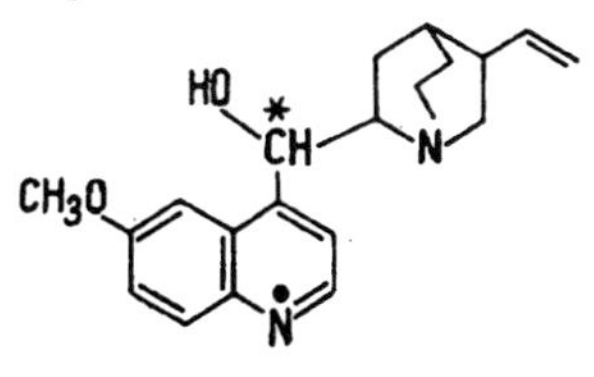			
Claviceps purpurea (Strain SD 58)				
DL-tryptophan-(alanine-3-^{14}C,-^{15}N)	elymoclavine	35.4	15.6	10
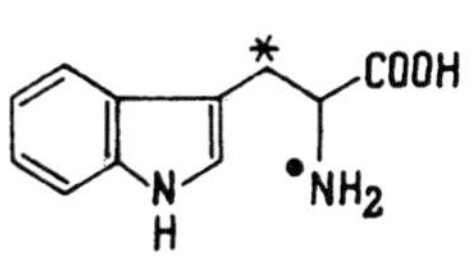	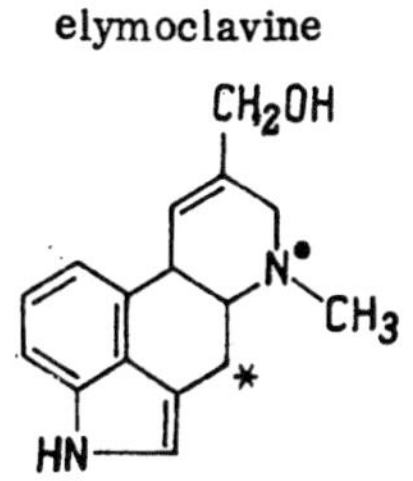			
D-tryptophan-	elymoclavine	7.1	0,7	10
(alanine-3-^{14}C-^{15}N)	elymoclavine	72.1	10.2	10

Species Precursor	Alkaloid	Specific % Incorporation ^{14}C	 ^{15}N	Reference
Colchicum autumnale				
autumnaline-9-^{14}C, ^{15}N	colchicine	0.64	0.64	11
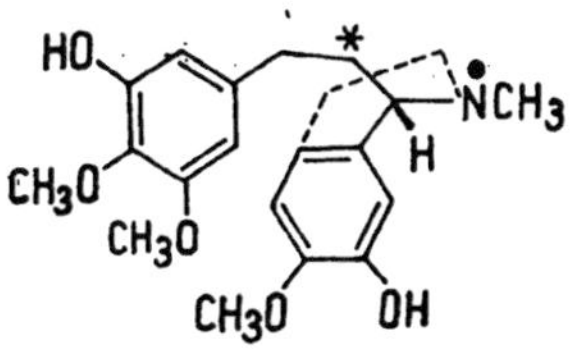	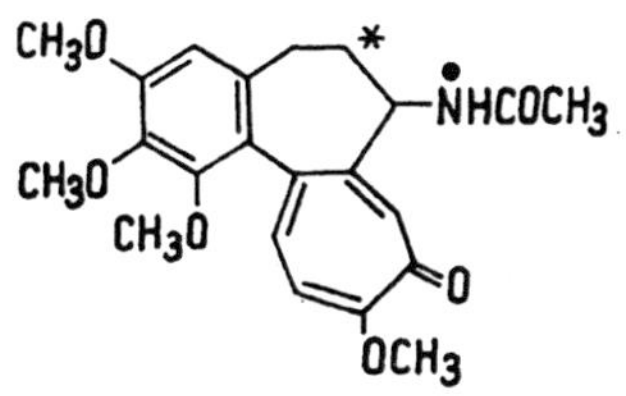			
Datura metel				
glutamic acid-2-^{14}C, ^{15}N	tropine	0.94	0.11	12
	scopoline	0.32	-	12
ornithine-2-^{14}C, δ-^{15}N	tropine	8.6	7.1	12
	scopoline	3.9	3.0	12
N-methylputrescine-methyl-^{14}C-^{15}N	hyoscyamine	3.03	2.81	13
$\overset{*}{C}H_3\overset{\bullet}{N}HCH_2CH_2CH_2CH_2NH_2$	hyoscine	0.56	0.51	13
Datura stramonium				
N-acetylglutamic acid-2-^{14}C-^{15}N	hyoscyamine	0.09	0.03	12
	hyoscine	0.07	-	12
	proline	0.13	0.05	12
ornithine-2-^{14}C, α-^{15}N	tropine	0.33	0.05	12
	scopoline	0.16	0.03	12
	proline	3.3	0.03	12
ornithine-2-^{14}C, δ-^{15}N	tropine	0.73	0.60	12
	scopoline	0.47	0.44	12
	proline	3.7	3.5	12
2-pyrrolidone-5-carboxylic acid-5-^{14}C, ^{15}N	hyoscyamine	0.04	0.03	12
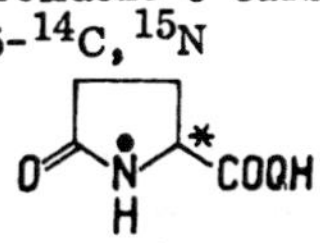	hyoscine	0.03	-	12

Species / Precursor	Alkaloid	Specific % Incorporation ^{14}C	Specific % Incorporation ^{15}N	Reference
Datura tatula				
ammonium sulfate-^{15}N	hyoscyamine	-	+	14
	hyoscine	-	+	14
Dictamnus albus				
anthranilic acid-carboxy-^{14}C, ^{15}N 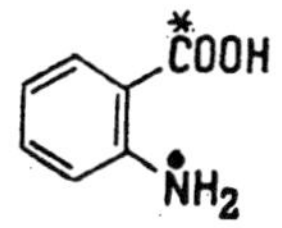	dictamnine 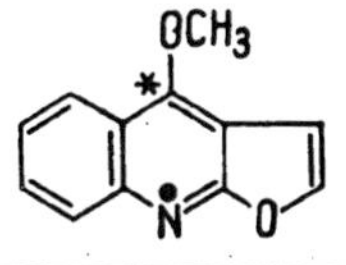	0.14	0.14	15
Hordeum vulgare				
tryptophan-(alanine-3^{14}C, ^{15}N)	gramine 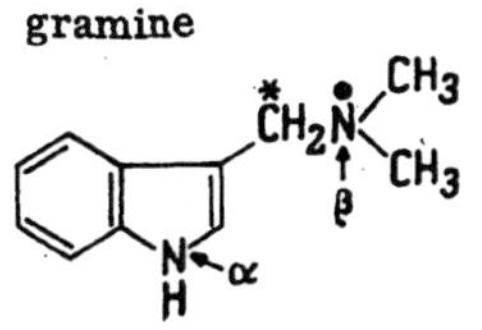	1.1	0.4 (all on N_β)	16
tryptophan-(alanine-^{15}N)	gramine	-	0.35	16
tryptophan + aspartic acid-^{15}N	gramine	-	0	16
tryptophan + glutamic acid-^{15}N	gramine	-	0	16
tryptophan + glutamine-amide-^{15}N	gramine	-	0	16
Lupinus luteus				
lysine-2-^{14}C,α-^{15}N 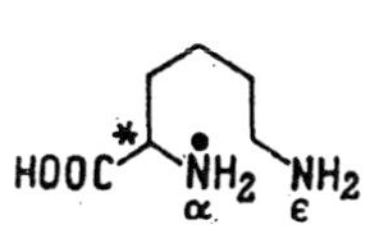	sparteine 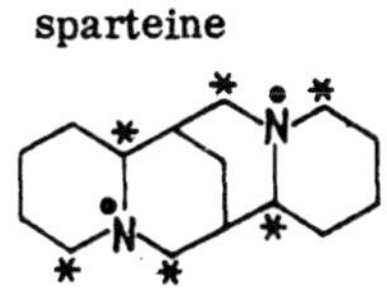	1.78	0.65	17
Mycobacterium tuberculosis				
aspartic acid-1,4-^{14}C-^{15}N 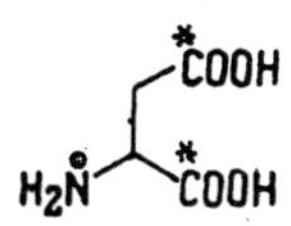	nicotinic acid 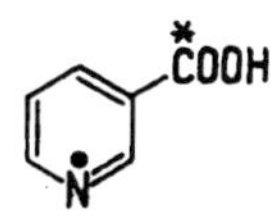	48	106	18
Nicotiana glauca				
lysine-2-^{14}C,α-^{15}N	anabasine 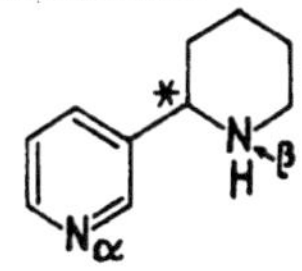	38.6	2.6	19

Species Precursor	Alkaloid	Specific % Incorporation ^{14}C	 ^{15}N	Reference
	nicotine 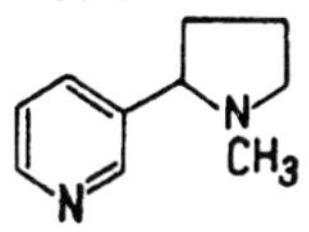	0.29	0.11	19
lysine-2-^{14}C, ϵ-^{15}N	anabasine	31.1	24.8 (on N_{β})	19
	nicotine	0.07	2.6	19
nicotine-^{15}N-methyl-^{14}C 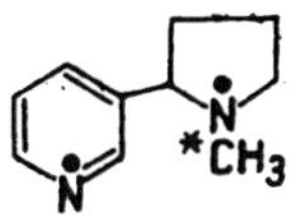	anabasine	4.4	2.2	20
	anatabine	4.9	6.4	20
	nicotine	21.2	36.6	20
	nornicotine	17.5	65.2	20
potassium nitrate-^{15}N	anabasine		+	21
Nicotiana glutinosa				
potassium nitrate-^{15}N	nornicotine		+	21
Nicotiana rustica				
anabasine-^{15}N	anabasine	-	56.0	20
	anatabine	-	6.5	20
	nicotine	-	4.0	20
	nornicotine	-	6.6	20
N-methylputrescine-(methyl-^{14}C, ^{15}N)	nicotine	0.62	0.64	22
N-methylputrescine-1-^{14}C, α-^{15}N	nicotine	0.26	0.28	23
nicotine-U-^{14}C, ^{15}N	nicotine* (* 83 % nicotine with minor alkaloids)	12.4	9.1	24
nicotine-^{15}N	anabasine	-	2.1	20
	anatabine	-	2.2	20
	nicotine	-	10.4	20
	nornicotine	-	2.7	20
nicotinic acid-^{15}N	nicotine (94 % of excess ^{15}N in pyridine ring)	-	+	25
nornicotine-^{15}N	anabasine	-	71.9	20
	anatabine	-	3.8	20
	nicotine	-	4.9	20
	nornicotine	-	40.5	20

Species Precursor	Alkaloid	Specific % Incorporation ^{14}C	 ^{15}N	Reference
potassium nitrate-^{15}N	nicotine	-	+	21
nitrate-^{15}N	nicotine	-	4-5.4	26
Nicotiana tabacum				
aspartic acid-^{15}N	nicotine (4 % absolute incorporation into both N)	-	+	27
6-hydroxynicotinic acid-^{15}N	nicotine	-	negl.	28
L-lysine-ϵ-^{15}N	nicotine 25 % on pyridine 75 % on pyrrolidine	-	0.43	29
nitrate-^{15}N	nicotine	-	+	21
	nornicotine	-	+	21
ornithine-2-^{14}C,α-^{15}N	nicotine	37	1.5	30
ornithine-2-^{14}C,δ-^{15}N	nicotine	61	31	30
Peganum harmala				
ornithine-α-^{15}N	peganine	-	0.24	31
	vasicinone 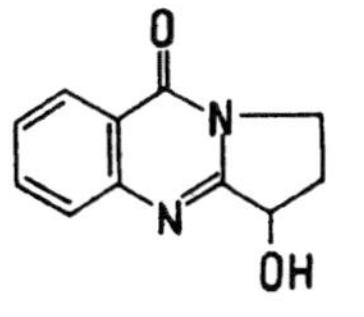	-	0.10	31
	desoxyvasicinone 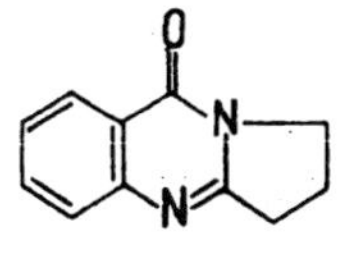	-	1.25	31
ornithine-δ-^{15}N	peganine	-	0.29	31
	vasicinone	-	0.12	31
	desoxyvasicinone	-	1.25	31
tryptophan-(alanine 3-^{14}C, ^{15}N)	harmine 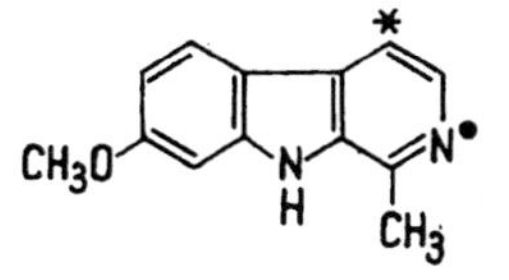	0.23	0.27	32
tryptamine-β-^{14}C, $^{15}N_{\beta}$	harmine	0.4	0.45	32

Species Precursor	Alkaloid	Specific % Incorporation ^{14}C	 ^{15}N	Reference
Phaseolus vulgaris				
lysine-2-^{14}C,α-^{15}N	pipecolic acid 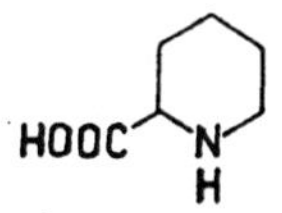	0.18	0.15	33
lysine-2-^{14}C,ε-^{15}N	pipecolic acid	0.42	0.10	33
Ricinus communis				
L-aspartic acid-^{15}N	ricinine 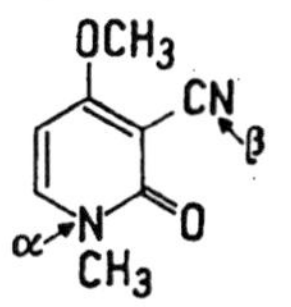	-	0.7-0.83 (50 % on each N)	34
DL-aspartic acid-4-^{14}C, ^{15}N	ricinine	0.38	Nα 0.20 Nβ 0.13	35
β-alanine-3-^{14}C, ^{15}N	ricinine	0.04	Nα 0.28 Nβ 0.08	36
N-methyl-β-alanine-methyl, 3-^{14}C, ^{15}N	ricinine	0.003	Nα 0.014 Nβ 0.038	36
potassium nitrate-^{15}N	ricinine	-	Nα 0.13 Nβ 0.27	37
ammonium nitrate-ammonium-^{15}N	ricinine	-	Nα 0.48 Nβ 0.94	37
formamide-^{15}N	ricinine	-	Nα 5.2 Nβ 7.3	37
L-glutamine-amide-^{15}N	ricinine	-	Nα 0.12 Nβ 0.50	37
nicotinamide-amide-^{15}N	ricinine	-	Nα 0.00 Nβ 0.58	37
nicotinamide-amide-^{15}N	ricinine	-	Nβ 1.36	38
nicotinamide-U-^{3}H-amide-^{15}N	ricinine	^{3}H:0.27	Nβ 1.29	38
1-methylnicotinamide-amide-^{15}N	ricinine	-	Nβ 0.10	37
nicotinonitrile-nitrile-^{15}N	ricinine	-	Nβ 0.06	37
1-methylnicotino-nitrile-nitrile-^{15}N	ricinine	-	Nβ 0.06	37

Species Precursor	Alkaloid	Specific % Incorporation ^{14}C	 ^{15}N	Reference
Sarothamnus scoparius				
lysine-2-^{14}C,α-^{15}N	sparteine	5.15	1.75	17
Trichoderma viride				
glycine-^{15}N	gliotoxin 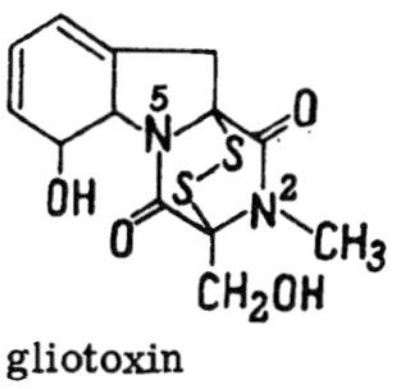	-	2.8-8.9 (on both N)	3
phenylalanine-^{15}N	gliotoxin	-	0.6-4.0	3
phenylalanine-1-^{14}C-^{15}N	gliotoxin	22.7	N_5 9.0 N_2 0	39
aspartic acid-1-^{14}C-^{15}N	gliotoxin	0	2.5	39
glutamic acid-^{15}N	gliotoxin	-	2.5	39

Specific incorporation is defined as follows:

$$\text{Specific incorporation} = \frac{\text{d.p.m./mM. of } ^{14}\text{C in the alkaloid}}{\text{d.p.m./mM. of } ^{14}\text{C in the administered precursor}}$$

$$= \frac{\%\text{ excess } ^{15}\text{N in the alkaloid}}{\%\text{ excess } ^{15}\text{N in the administered precursor}}$$

The location of the ^{14}C and the ^{15}N in the administered precursor and the resultant alkaloids are indicated in the table by ∗ (^{14}C) and •(^{15}N).

Experiments with ^{15}N labelled α-amino acids must be interpreted with caution because of the high propability of the ^{15}N entering the general nitrogen pool by a transamination. It is advisable in all cases to have ^{14}C present at some position in the α-amino acid, as an internal control, so that the amount of transamination which takes place can be estimated. Thus the results of Liljegren (31) on the biosynthesis of peganine in Peganum harmala are open to question. He fed ornithine -α-^{15}N and -δ-^{15}N and obtained incorporation into peganine. He was thus led to suggest that ornithine was incorporated, perhaps via putrescine, into the pyrrolidine ring of peganine. However Johne and Gröger (40) found that the administration of ornithine-2-^{14}C or putrescine-1,4-^{14}C to Adhatoda vasica plants yielded radioactive peganine which was not specifically labelled in the pyrrolidine ring.

The incorporation of aspartic acid-^{15}N into nicotinic acid (18), or its metabolites : nicotine (27) and ricinine (34, 35) also must be interpreted with caution because of the rapid equilibration of the amino group of aspartic acid with the nitrogen pool. Work with isolated enzymes is required to definitely establish aspartic acid as a direct precursor of nicotinic acid in higher plants. In the bacteria Clostridium butylicum nicotinic acid is formed from aspartic acid, and a cell free system has been obtained (41) which converts N-formyl-aspartic acid to nicotinic acid. Tracer work with intact cells of this bacteria has indicated that C_4 and C_5 of nicotinic acid are derived from acetic acid (42). A plausible biosynthetic scheme is illustrated below in Fig. 1.

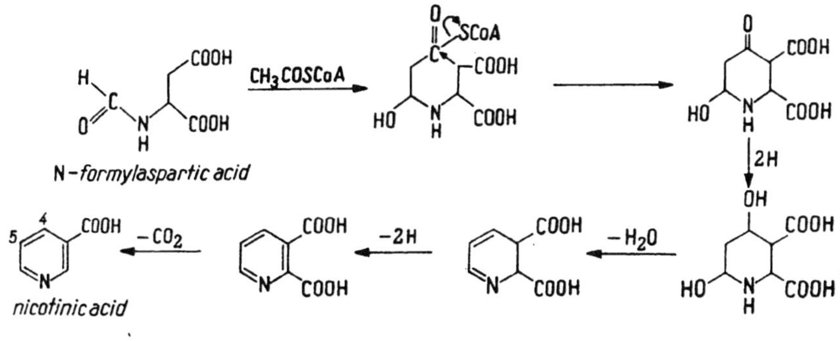

Figure 1. Hypothetical formation of nicotinic acid in Clostridium butylicum

It is not possible to generalize about the origin of piperidine rings. At least two distinct pathways exist for the formation of the piperidine ring in higher plants (see Fig. 2). One route is from lysine, and this is the precursor of the piperidine ring of anabasine (19), pipecolic acid (33, and references cited therin), N-methyl-pelletierine (43, 44), sedamine (45), and lycopodine (46). The incorporation of lysine into these alkaloids does not apparently proceed via a symmetrical intermediate such as cadaverine, since the administration of lysine-2-^{14}C or -6-^{14}C afforded alkaloids which were labelled at only one of the α-positions of the piperidine rings (19, 44). Δ'-Piperideine was shown to be the immediate precursor of the piperidine of anabasine (47), Δ'-piperideine-6-^{14}C affording anabasine-6'-^{14}C when fed to Nicotiana glauca plants. A second route to the piperidine ring is by the cyclization of a polyketide chain derived from acetic acid. This route has been established for the Hemlock alkaloids, coniine, conhydrine and Ψ-conhydrine (48, 49) and for pinidine in Pinus jeffreyi (50). The source of the nitrogen in these alkaloids has not been established.

Although no ^{15}N tracer work has been carried out on the isoquinoline alkaloids (except on colchicine, which is a modified phenethyl-isoquinoline alkaloid (11)), the evidence from ^{14}C studies is so overwhelming, one can safely assume that the nitrogen in these alkaloids is derived from the amino group of dopamine or a related phenethylamine (51, 52).

It also seems highly probable that the two nitrogens found in so many indole alkaloids are derived from the two nitrogens of tryptophan. However ^{15}N studies with labelled tryptophan would be worthwhile in the case of alkaloids such as uleine and ellipticine, which have apparently lost the two carbon bridge joining the β-position of the indole nucleus with the α-amino group of tryptophan.

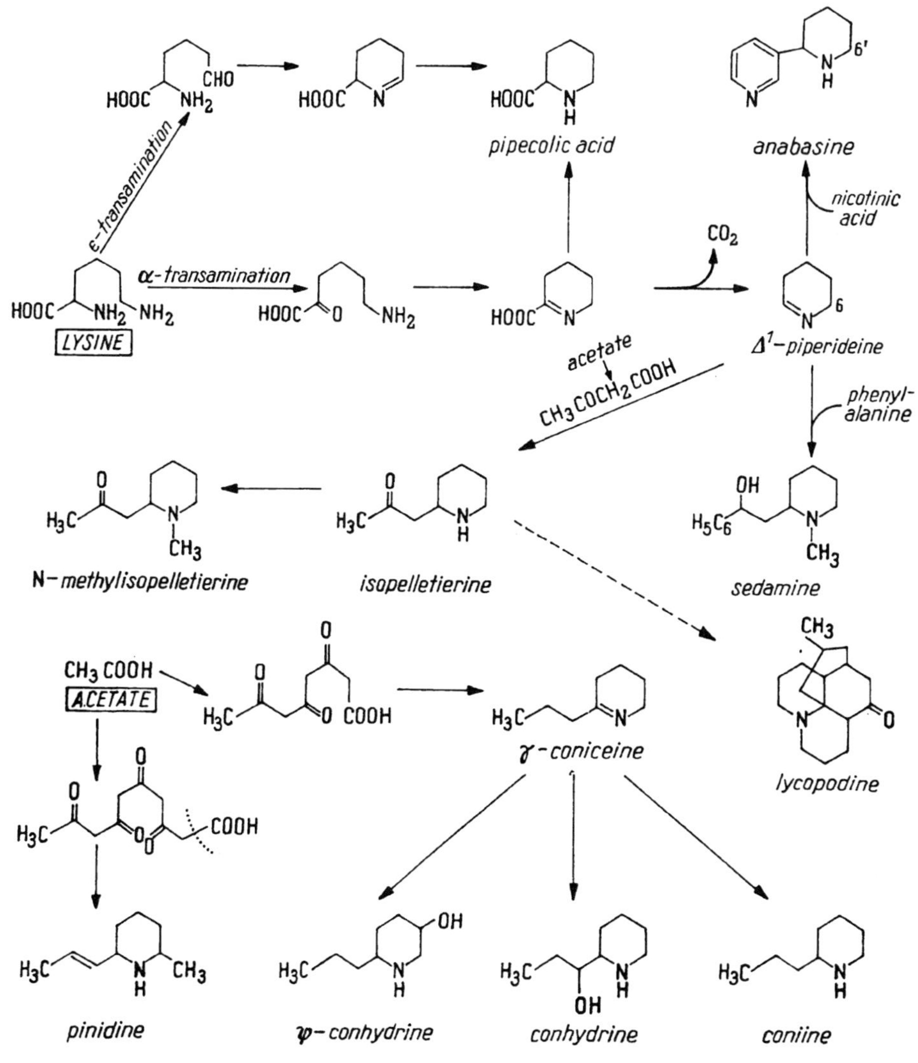

Figure 2. Some biosynthetic routes to Piperidine rings

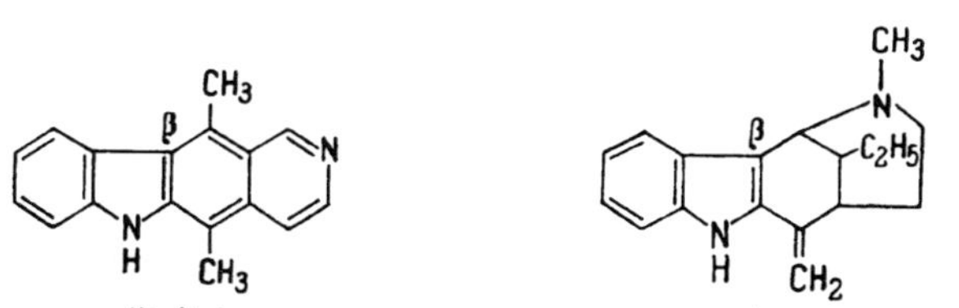

ellipticine *uleine*

Some other alkaloids in which the origin of the nitrogen is in doubt, and has been the subject of speculation, are illustrated in Fig. 3, together with tentative biogenetic schemes. Tracer studies with ^{15}N labelled precursors should resolve these problems in the near future.

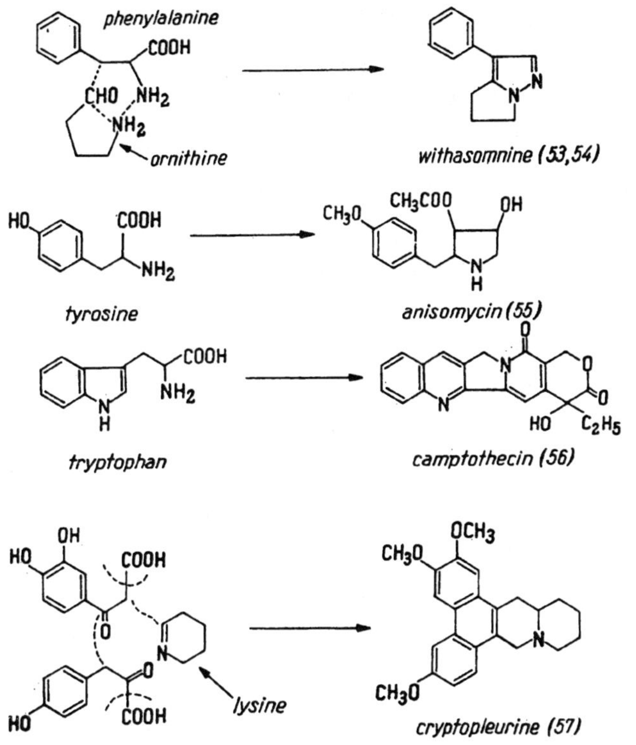

Figure 3. Tentative biogenetic schemes for some alkaloids (references in parentheses)

References

(1) Günther, H., H.G. Floss, and H. Simon, Z. analyt. Chem. 218, 401 (1966)

(2) Waller, G.R., R. Ryhage, and S. Meyerson, Analyt. Biochem. (New York) 16, 277 (1966)

(3) Bose, A.K., K.G. Das, P.T. Funke, I. Kugajevsky, O.P. Shulkla, K.S. Khanchandani, and R.J. Suhadolnik, J. Amer. chem. Soc. 90, 1038 (1968)

(4) Faust, H., Z. analyt. Chem. 175, 9 (1960)

(5) Meier, G., Kernenergie 3, 903 (1960)

(6) Munsche, D., Isotopenpraxis 1, 32 (1965)

(7) Johne, S., D. Gröger, and G. Richter, Arch. Pharmaz. 301, 721 (1968)

(8) Leete, E., and E.G. Gros, Abh. Dtsch. Akad. Wiss. Berlin, Kl. Chem., Geol., Biol., Jg. 1966, Nr. 3, p 538, Akademie-Verlag, Berlin 1966

(9) Leete, E., and J.N. Wemple, J. Amer. chem. Soc. 91 (April) (1969)

(10) Floss, H.G., U. Mothes, and H. Günther, Z. Naturforsch. 19b, 784 (1964)

(11) Battersby, A.R., R.B. Herbert, E. McDonald, R. Ramage, and J.H. Clements, Chem. Commun. 603 (1966)

(12) Liebisch, H.W., and H.R. Schütte, Z. Pflanzenphysiol. 57, 434 (1967)

(13) Liebisch, H.W., W. Maier, and H.R. Schütte, Tetrahedron Letters (London) 1966, 4079

(14) Imaseki, I., Pharmac. Bull. (Tokyo) 3, 328 (1955); 5, 447 (1957)

(15) Monkovic, I., I.D. Spenser, and A.O. Plunkett, Canad. J. Chem. 45, 1935 (1967)

(16) Gross, D., A. Nemeckova, and H.R. Schütte, Z. Pflanzenphysiol. 57, 60 (1967)

(17) Schütte, H.R., H. Hindorf, K. Mothes, and G. Hübner, Liebigs Ann. Chem. 680, 93 (1964)

(18) Gross, D., H.R. Schütte, G. Hübner, and K. Mothes, Tetrahedron Letters (London), 1963, 541

(19) Leete, E., E.G. Gross, and T.J. Gilbertson, J. Amer. chem. Soc. 86, 3907 (1964)

(20) Tso, T.C., and R.N. Jeffrey, Arch. Biochem. Biophysics 80, 46 (1959)

(21) Tso, T.C., and R.N. Jeffrey, Plant Physiol. 32, 86 (1957)

(22) Schütte, H.R., W. Maier, and K. Mothes, Acta Biochim. Polon. 13, 401 (1966)

(23) Schütte, H.R., W. Maier, and U. Stephan, Z. Naturforsch. 23b, 1426 (1968)

(24) Tso, T.C., and R.N. Jeffrey, Arch. Biochem. Biophysics 92, 253 (1961)

(25) Tso, T.C., and R.N. Jeffrey "The formation and interconversion of ^{15}N labelled alkaloids in tobacco plants", Tobacco Chemists Research Conference, Washington D.C. 1956

(26) Tso, T.C., Phytochemistry 5, 287 (1966)

(27) Christman, D.R., and R.F. Dawson, Biochemistry 2, 182 (1963)

(28) Dawson, R.F., D.R. Christman, A. D'Adamo, and A.P. Wolf, J. Amer. chem. Soc. 82, 2628 (1960)

(29) Bothner-By, A.A., R.F. Dawson, and D.R. Christman, Experientia 12, 151 (1956)

(30) Leete, E., E.G. Gros, and T.J. Gilbertson, Tetrahedron Letters (London), 1964, 587

(31) Liljegren, D.R., Phytochemistry 7, 1299 (1968)

(32) Stolle, K., and D. Gröger, Arch. Pharmaz. 301, 561 (1968)

(33) Schütte, H.R., and G. Seelig, Z. Naturforsch. 22b, 824 (1967)

(34) Johns, S.R., and L. Marion, Canad. J. Chem. 44, 23 (1966)

(35) Yang, K.S., and G.R. Waller, Phytochemistry 4, 881 (1965)

(36) Gross, D., P. Banditt, J.W. Kurbatow, and H.R. Schütte, Z. Pflanzenphysiol. 58, 410 (1968)

(37) Waller, G.R., and K.S. Yang, Phytochemistry 6, 1637 (1967)

(38) Waller, G.R., and L.M. Henderson, J. biol. Chemistry 236, 1186 (1961)

(39) Bose, A.K., K.S. Khanchandani, R. Tavares, and P.T. Funke, J. Amer. chem. Soc. 90, 3594 (1968)

(40) Johne, S., and D. Gröger, Phytochemistry 7, 429 (1968)

(41) Scott, T.A., E. Bellion, and M. Mattey, Biochem. J. 107, 23p (1968)

(42) Bellion, E., Ph. D. thesis, Department of Biochemistry, University of Leeds 1968

(43) O'Donovan, D.G., and M.F. Keogh, Tetrahedron Letters (London) 1968, 265

(44) Gupta, R.N., and I.D. Spenser, Chem. Commun. 1968, 85

(45) Gupta, R.N., and I.D. Spenser, Chem. Commun. 1966, 893; Canad. J. Chem. 45, 1275 (1967)

(46) Gupta, R.N., M. Castillo, D.B. McLean, I.D. Spenser, and J.T. Wrobel, J. Amer. chem. Soc. 90, 1360 (1968)

(47) Leete, E., J. Amer. chem. Soc. 91, in press (1969)

(48) Leete, E., J. Amer. chem. Soc. 85, 3523 (1963); 86, 2509 (1964)

(49) Leete, E., and N. Adityachaudhury, Phytochemistry 6, 219 (1967)

(50) Leete, E., and K.N. Juneau, J. Amer. chem. Soc., in press (1969)

(51) Barton, D.H.R., Chem. in Britain 330 (1967)

(52) Spenser, I.D., Lloydia 29, 71 (1966)

(53) Onaka, T., Tetrahedron Letters (London) 1968, 5711

(54) Schröter, H.B., D. Neumann, A.R. Katritzky, and F.J. Swinbourne, Tetrahedron (London) 22, 2895 (1966)

(55) Butler, K., J. org. Chemistry 31, 317 (1966)

(56) Wenkert, E., K.G. Dave, R.G. Lewis, and P.W. Sprague, J. Amer. chem. Soc. 89, 6741 (1967)

(57) Wenkert, E., Experientia (Basel) 15, 165 (1959)

(Aus dem Institut für Mikrobiologie und Molekularbiologie
der Universität Hohenheim, DBR)

Über Regulationsmechanismen bei der Biosynthese von Alkaloidvorstufen

Von F. Lingens

Als Ausgangsmaterial für die Biosynthese von Alkaloiden dienen vielfach Aminosäuren. Diese Alkaloidvorstufen sind gleichzeitig Endprodukte des anabolischen Stoffwechsels. Die Biosynthese der Aminosäuren wird im allgemeinen so gesteuert, daß die Endprodukte der Syntheseketten nicht im Überschuß gebildet werden. Dies kann dadurch geschehen, daß das Endprodukt die Biosynthese der Enzyme der ganzen Kette verhindert. Dieser Vorgang wird als Repression bezeichnet. Der erste Umsetzungsschritt wird normalerweise von einem Enzym katalysiert, das durch das Endprodukt der Kette in seiner Aktivität gehemmt wird. Bei einem solchen hemmbaren Enzym handelt es sich im allgemeinen um ein allosterisches Enzym.

Diese allosterischen Proteine sind nach Monod, Wyman und Changeux (1) Oligomere aus identischen Protomeren, die durch nicht-kovalente Bindungen verknüpft sind. Jedes Protomere besitzt stereospezifische Stellen für das Substrat und andere Liganden (Effektoren). Das Oligomere (allosterisches Protein) liegt in den energetisch verschiedenen Zuständen R und T vor, die miteinander im Gleichgewicht stehen. R- und T-Zustand unterscheiden sich in der Quartärstruktur des Proteins. Die Liganden haben verschiedene Affinität zu den beiden Zuständen. Das Gleichgewicht zwischen R- und T-Zustand hängt von der Konzentration der Liganden ab. Dieses Modell ist in zwei Formen realisiert:

1. das K-System: Die Effektoren und das Substrat haben verschiedene Affinitäten zum R- und T-Zustand. Dadurch beeinflußt der Effektor die Affinität des Enzymproteins zum Substrat, das Substrat die Affinität des Proteins zum Effektor.
2. das V-System: Hier hat das Substrat die gleiche Affinität zu beiden Zuständen. Dadurch hat ein Effektor keinen Einfluß auf die Anlagerung des Substrats und umgekehrt. Die beiden Zustände unterscheiden sich aber in ihrer katalytischen Aktivität, so daß ein Effektor, der das Gleichgewicht zugunsten des Zustandes mit der niedrigeren katalytischen Aktivität verschiebt, das Enzym hemmt.

Bei einer verzweigten Biosynthesekette, wie sie z. B. für die Biosynthese der aromatischen Aminosäuren (Abb. 1) gegeben ist, wird die Regulation dadurch kompliziert, daß nicht nur jede einzelne Kette, sondern auch der Grundzweig einer Regulation unterworfen werden sollte. Der erste Schritt des Grundzweiges kann durch Isoenzyme katalysiert sein, von denen jedes durch ein hinter der Verzweigung liegendes Endprodukt spezifisch allosterisch gehemmt und (oder) reprimiert wird. Es ist aber auch denkbar, daß für den ersten Schritt des Grundzweiges nur ein einziges Enzym vorliegt, das entweder durch das Endprodukt des Grundzweiges oder durch die Summe der Endprodukte allosterisch gehemmt oder reprimiert wird. Bei einer verzweigten Biosynthesekette ist weiterhin zu beachten, daß das Endprodukt einer Kette nicht nur hemmend auf die eigene Kette einwirken kann, sondern auch einen positiven Effekt durch Induktion oder allosterische Aktivierung auf einen anderen Zweig ausüben kann.

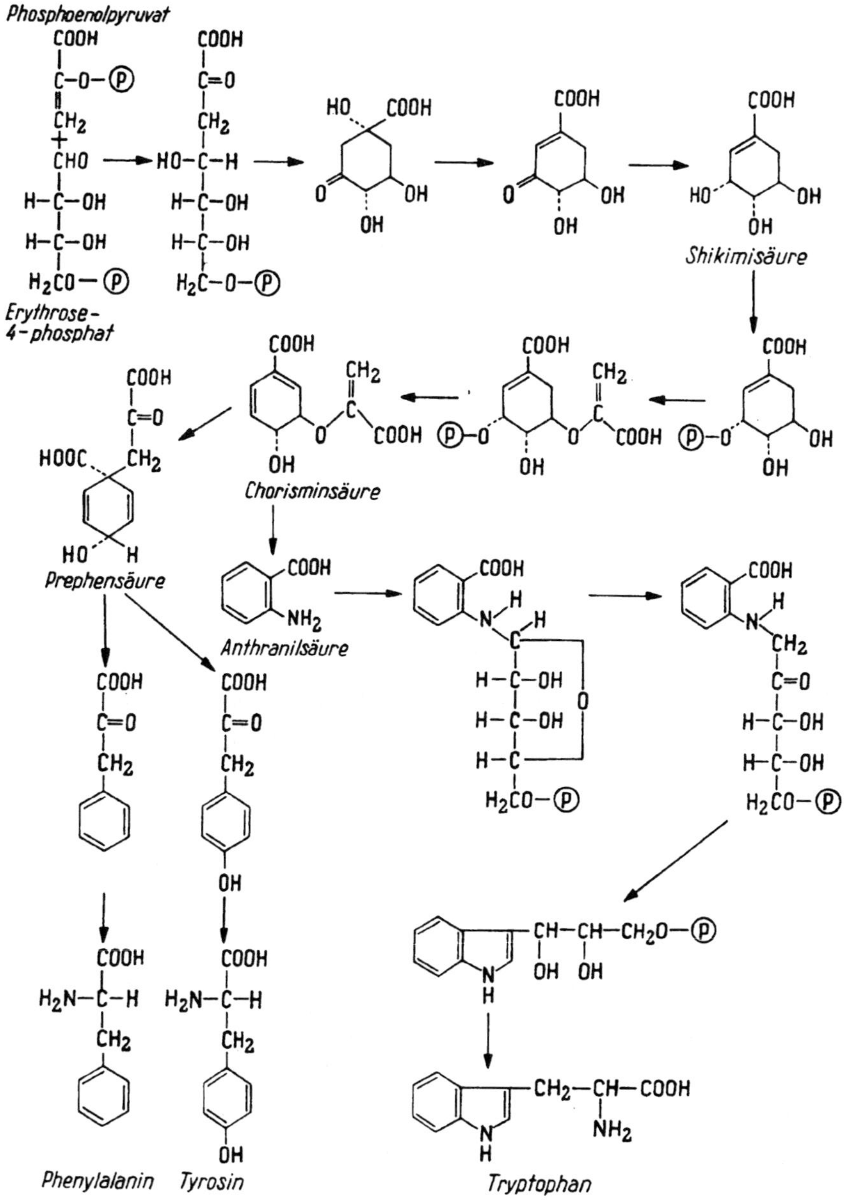

Abbildung 1.

Es ist erwiesen, daß L-Tryptophan eines der Ausgangsprodukte für die Biosynthese der Mutterkornalkaloide ist. Andererseits ist L-Tryptophan eines der Endprodukte der verzweigten Biosyntheseketten der aromatischen Aminosäuren. Deshalb schien eine Untersuchung der Regulation der Biosynthese dieser aromatischen Aminosäuren im Zusammenhang mit dem Problem der Mutterkornalkaloidbildung nicht uninteressant. Für die Regulation dieser Biosynthese sind vor allem die Eigenschaften der folgenden Enzyme zu betrachten: 2-Keto-3-desoxy-araboheptonsäure-7-phosphat-Synthetase (DAHP-Synthetase) (katalysiert die Kondensation von Phosphoenolbrenztraubensäure mit Erythrose-4-phosphat zu 2-Keto-3-desoxy-arabo-heptonsäure-7-phosphat), Chorismat-Mutase (katalysiert die Umlagerung von Chorisminsäure in Prephensäure), Prephenat-Dehydrogenase (bewirkt die Umwandlung von Prephensäure in p-Hydroxyphenylbrenztraubensäure), Prephenat-Dehydratase (bewirkt die Umwandlung von Prephensäure in Phenylbrenztraubensäure) und Anthranilat-Synthetase (katalysiert die Umwandlung von Chorisminsäure zu Anthranilsäure) (Abb. 2).

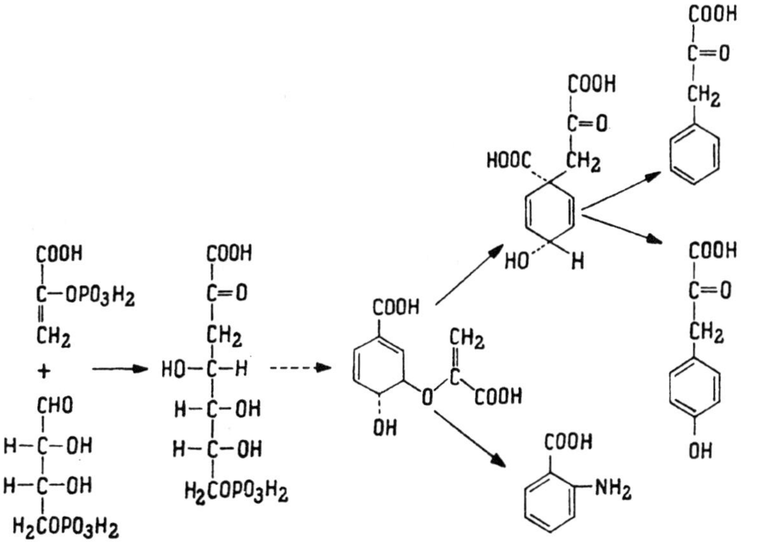

Abbildung 2.

Bei der Untersuchung des Alkaloid produzierenden Claviceps paspali-Stammes PB 156 - E 1/6 stellte es sich heraus, daß die DAHP-Synthetaseaktivität dieses Stammes sich durch Fraktionierung mit Ammoniumsulfat in drei Isoenzyme auftrennen läßt (2). In der Fraktion 0 - 0,35 gesättigt mit Ammoniumsulfat erscheint das tryptophansensitive Isoenzym, dessen Aktivität durch L-Tryptophan hemmbar ist und ca. 60 % der Gesamt-DAHP-Synthetaseaktivität dieses Organismus ausmacht. In der Ammoniumsulfatfraktion 0,35 - 0,45 gesättigt findet man das durch L-Tyrosin hemmbare Isoenzym, in der Fraktion 0,45 - 0,5 gesättigt mit Ammoniumsulfat das durch L-Phenylalanin hemmbare Isoenzym. Die Stabilität der beiden letztgenannten Isoenzyme ist wesentlich geringer als die des tryptophansensitiven Isoenzyms.

Die Zusammensetzung der Gesamt-DAHP-Synthetaseaktivität ist hier also ganz anders als bei Escherichia coli und Saccharomyces cerevisiae, deren Gesamtaktivität an DAHP-Synthetase hauptsächlich aus den beiden phenylalanin- und tyrosinsensitiven Isoenzymen besteht. Eine vorhandene

Restaktivität ist durch L-Tryptophan nicht hemmbar (2). Die Enzyme Chorismat-Mutase, Prephenat-Dehydratase und Prephenat-Dehydrogenase aus Claviceps paspali, deren Aktivität jeweils durch die Endprodukte (Tyrosin, Phenylalanin) der betreffenden Synthesekette hemmbar ist, unterscheiden sich nicht wesentlich von den entsprechenden Enzymen aus E. coli und S. cerevisiae (2, 3).

Die Untersuchungen an DAHP-Synthetase und Chorismat-Mutase bei Claviceps purpurea SD 58 lieferten Ergebnisse, die weitgehend denen von Claviceps paspali PB 156 - E 1/6 entsprechen. Die Isoenzyme der DAHP-Synthetase sind allerdings erst bei etwas höherer Sättigung mit Ammoniumsulfat fällbar. Die DAHP-Synthetase benötigt offenbar Co^{++} als Cofaktor, da die spezifische Aktivität im Rohextrakt durch Behandlung mit EDTA völlig verloren geht, bei nachfolgender Behandlung mit Co^{++} aber auf den Ausgangswert restauriert werden kann. Durch Säulenchromatographie an DEAE-Cellulose läßt sich das tryptophansensitive Isoenzym der DAHP-Synthetase von den beiden anderen Isoenzymen abtrennen (4).

Im folgenden soll näher auf die Chorismat-Mutase eingegangen werden, da dieses Enzym besonders interessante Eigenschaften aufweist.

Die Chorismat-Mutase kanalisiert die Chorisminsäure in die Richtung L-Phenylalanin und L-Tyrosin und kann dadurch den Umsatz der Anthranilat-Synthetase in Richtung L-Tryptophan und damit eventuell auch die Alkaloidausbeute begrenzen. Die Chorismatmutase ist als allosterisches Enzym dem K-System zuzuordnen. Die Aktivität des Enzyms läßt sich durch L-Phenylalanin und L-Tyrosin hemmen und schon durch geringe Mengen L-Tryptophan aktivieren. Diese Aktivierung kann zum enzymatischen Nachweis geringer Mengen L-Tryptophan verwendet werden. Das pH-Optimum des Enzyms liegt bei pH 6,6, bei pH 8 findet in Abwesenheit von L-Tryptophan kein, in Anwesenheit von L-Tryptophan ein großer enzymatischer Umsatz statt. L-Tryptophan stabilisiert das Enzym, so daß es bei 30° C einige Tage voll aktiv bleibt. Weniger gut wirken einige Analoge des L-Tryptophans (5).

L-Tryptophan hebt die Hemmung der Enzymaktivität durch L-Phenylalanin und L-Tyrosin (10^{-3} m) schon in geringen Konzentrationen (10^{-5} m) auf und bewirkt darüber hinaus bei höherer Konzentration Aktivierung. Durch Chromatographie an DEAE-Cellulose kann das Enzym 40-fach angereichert werden. Diese gereinigte Chorismat-Mutase verliert aber leicht die Hemmbarkeit durch L-Phenylalanin und L-Tyrosin bei vollem Erhalt der Aktivierbarkeit durch L-Tryptophan, wenn dem Puffer nicht Mercaptoäthanol (10^{-2} m) zugesetzt worden ist. Dieses nicht mehr hemmbare Enzym zeigt in Anwesenheit von L-Phenylalanin und L-Tyrosin sogar einen schwach erhöhten enzymatischen Umsatz. Die Chorismat-Mutase ist nicht nur durch L-Tryptophan, sondern auch durch viele Analoge des L-Tryptophans aktivierbar, die allerdings für eine entsprechende Aktivierung in höherer Konzentration vorliegen müssen. Zum Verlust dieser Aktivierungsfähigkeit führen Substitutionen in 7-Stellung (7-Methyl- bzw. 7-Äthyl-DL-tryptophan), größere Reste in 4,5- und 5,6-Stellung (Benz-DL-tryptophane), Änderungen in der Länge der Seitenkette, Ersatz des Indolstickstoffs durch Schwefel, Vergrößerung des 5-Rings. Im übrigen ist für Substitutionen, die nicht zu völligem Verlust der Aktivierungsfähigkeit führen, ein breiter Spielraum gegeben (Methyl-Substitution, Änderung der Zahl und Lage der Stickstoffatome im Indolkern). Interessant ist weiterhin, daß auch L-Histidin und L-Cystein (Aminosäuren anderer Biosyntheseketten) und EDTA aktivierend auf die Chorismat-Mutase wirken (5).

Das für die Alkaloidbiosynthese benötigte L-Tryptophan kann entweder aus den Proteinen freigesetzt oder auf dem oben gezeigten Biosyntheseweg aufgebaut werden (6). Wir haben festgestellt, daß noch lange nach Beendigung des Wachstums die von uns untersuchten Enzyme der Biosynthese aromatischer Aminosäuren vorhanden und sehr aktiv sind. Außerdem konnte von Rothe und Fritsche durch Einbauversuche mit radioaktivem Schwefel gezeigt werden, daß beim Claviceps Stamm SD 58 auch nach Abschluß des Wachstums noch eine Eiweißsynthese stattfindet (7). Wir dürfen also annehmen, daß das für die Alkaloidbiosynthese benötigte L-Tryptophan mindestens zum Teil durch de novo-Synthese zur Verfügung steht. Bei dem sehr gut Alkaloid-produzierenden Sandoz-Stamm PB 156 - E 1/6 fanden wir, daß die Aktivität des ersten Enzyms der eigentlichen Tryptophanbiosynthese, die Anthranilat-Synthetase, durch L-Tryptophan nicht verringert wird. Die Frage bleibt bis jetzt allerdings offen, ob dieser Stamm allein aus diesem Grunde vermehrt Alkaloid bildet, da ein Überschuß an L-Tryptophan im Medium nicht in jedem Falle eine Steigerung der Alkaloidbiosynthese zur Folge haben muß. So konnte beispielsweise bei dem Stamm Li 189 durch appliziertes L-Tryptophan keine Steigerung der Alkaloidausbeute erzielt werden (8), obwohl der Stamm SD 58 unter diesen Bedingungen erhöhte Alkaloidproduktion zeigte (9). Anscheinend wird das L-Tryptophan in dem erstgenannten Falle schneller in Richtung auf Acetylkynurenin abgebaut. Es gibt noch einige andere Abbaumöglichkeiten für L-Tryptophan; der eingeschlagene Weg scheint wesentlich von der Zusammensetzung der Nährlösung abzuhängen (9), so daß auch auf Grund der Vielseitigkeit des Tryptophanstoffwechsels für eine optimale Alkaloidsynthese ganz bestimmte Züchtungsbedingungen eingehalten werden müssen.

Die Enzym-Regulation scheint ebenfalls nicht bei allen Claviceps-Stämmen einheitlich verwirklicht zu sein. In Extrakten aus Mycel des Stammes SD 58 konnten wir nach anfänglichen Schwierigkeiten die Anthranilat-Synthetase nachweisen, die aber durch L-Tryptophan in ihrer Aktivität stark gehemmt werden konnte (10).

Die eigenen Untersuchungen entstanden in Zusammenarbeit mit Fräulein Dr. H. Uesseler und den Herren Dr. W. Goebel, Dr. B. Sprößler, Dipl. Chem. J. Eberspächer, Dr. med. I. Rahimi-Laridjani und Dipl. Chem. H. Schmid.

Dank: Herrn Dr. D. Gröger, Halle, danken wir für Überlassung mehrerer Claviceps-Stämme, der Sandoz-AG, Basel, danken wir für die Züchtung und Überlassung von Claviceps paspali-Mycel.

Literatur

(1) Monod, J., J. Wyman u. J.P. Changeux, J. molecular Biol. 12, 88 (1965)

(2) Lingens, F., W. Goebel u. H. Uesseler, Europ. J. Biochem. 2, 442 (1967)

(3) Lingens, F., W. Goebel u. H. Uesseler, Biochem. Z. 346, 357 (1966);
Europ. J. Biochem. 1, 363 (1967)

(4) Uesseler, H., J. Eberspächer u. F. Lingens, unveröffentlicht

(5) Sprößler, B. u. F. Lingens, unveröffentlicht

(6) Gröger, D., Pharmazie 15, 715 (1960)

(7) Rothe, U. u. W. Fritsche, Arch. Mikrobiol. 58, 77 (1968)

(8) Gröger, D. u. D. Erge, Pharmazie 19, 775 (1964)

(9) Gröger, D., Arch. Pharmaz. 292, 389 (1959)

(10) Sprößler, B., H. Schmid u. F. Lingens, unveröffentlicht

(Aus dem Organisch-chemischen Institut der Universität Zürich, Schweiz)

Bisindolalkaloide

H. Schmid

In den letzten Jahren hat sich - in struktureller Hinsicht betrachtet - besonders die Chemie der Bisindolalkaloide oder, wie sie häufig bezeichnet werden, der "dimeren" Indolalkaloide, stark entwickelt.*) Bei diesen Basen handelt es sich um Stoffe, die aus der Verbindung von zwei individuellen, eindeutig als Alkaloide im engeren Sinne anzusprechenden Teilen hervorgehen. Obwohl manche Bisindolalkaloide schon seit längerer Zeit bekannt waren, geht ihre Konstitutionsaufklärung auf die letzten 10 Jahre zurück. Daraus folgt, daß die modernen mikropräparativen und spektroskopischen Methoden (magnetische Kernresonanz, massenspektrometrische und Röntgenstrukturanalyse) geradezu eine Voraussetzung für erfolgreiche Strukturaufklärung waren.

Die Bisindolalkaloide lassen sich in zwei Gruppen einteilen: Die erste umfaßt Basen aus zwei gleichen oder zumindest sehr stark verwandten "Hälften", bei denen jeweils dieselben Zentren für die Verknüpfung Verwendung finden. Die Calycanthus**)- und insbesondere die Alkaloide aus Calebassencurare und südamerikanischen Strychnaceen gehören hierzu. Die Stoffe sind häufig durch Symmetrieelemente gekennzeichnet, das Skelett der (optisch aktiven) Curarealkaloide etwa durch eine zweizählige Symmetrieachse.

Die zweite Gruppe besteht aus Basen mit verschiedenen "monomeren" Alkaloidkomponenten (z.B. Geissospermin, Vinblastin, Villalstonin) oder bei den die beiden typgleichen Teile jeweils durch verschiedene Zentren miteinander verknüpft sind. Symmetrieeigenschaften gehen diesen Bisindolalkaloiden ab.

Häufig weisen die Bisindolbasen für die Gesamtstruktur charakteristische chemische und spektroskopische Eigenschaften auf. Von Interesse sind ferner pharmakologische Eigenschaften, wie die muskelrelaxierende Wirkung der bisquartären Curarealkaloide - klinisch wird das halbsynthetische Alloferin eingesetzt - sowie die ebenfalls klinisch verwendeten, antileukämisch wirksamen Vinca-Alkaloide Leurocristin und Vinblastin.

*) Angaben über Vorkommen, Strukturen etc. von Alkaloiden finden sich, soweit nicht besonders aufgeführt, in den Tabellen von M. Hesse (1).

**) Hierher gehört auch das "trimere" Hodgkinsin.

Bei den Curarealkaloiden - als Vertreter der ersten Gruppe - dürfen die Probleme der Strukturaufklärung als weitgehend gelöst betrachtet werden. Ungelöst ist noch eine Vielzahl chemischer Probleme. Alle "dimeren" Curarealkaloide leiten sich vom sogenannten Wieland-Gumlich-aldehyd (I) ab, den Wieland 1933 durch einen Abbau des (leicht zugänglichen) Strychnins (II) erstmals gewonnen hatte (2).

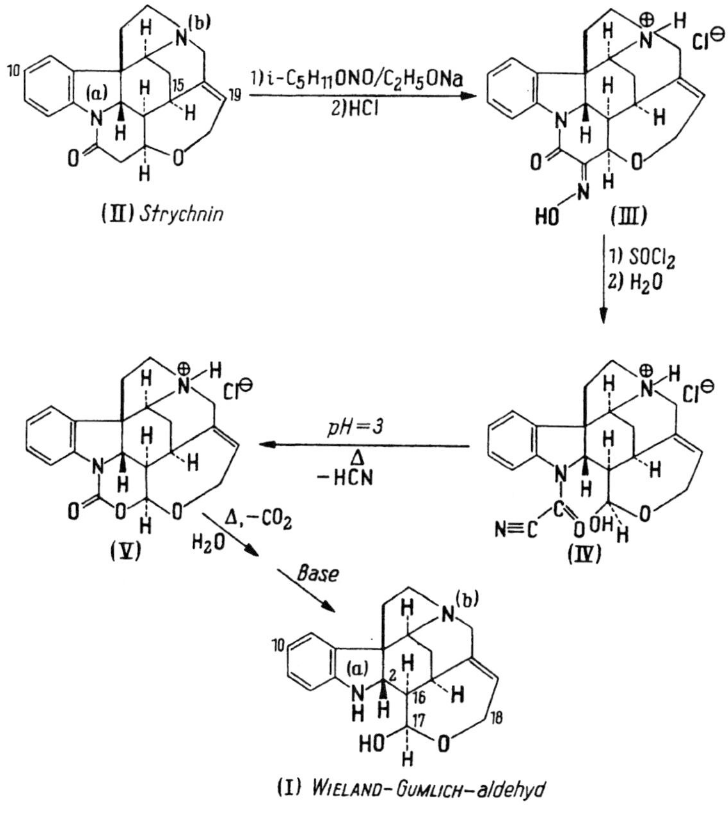

Der Verlauf dieses Abbaues ist erst kürzlich geklärt worden und hat u. a. zu einer Steigerung der Ausbeute an I geführt (3). Die wichtigsten Stufen sind nachstehend wiedergegeben:
Durch Isonitrosierung wird II zur Hauptsache in III umgewandelt, das mit Thionylchlorid eine Fragmentierung (Beckmann-Umlagerung II. Art) zu IV erleidet. Dieses Zwischenprodukt gibt beim Erwärmen in saurer, wässeriger Lösung zuerst Blausäure (unter Bildung von V) und spaltet hierauf CO_2 ab, wobei Wieland-Gumlich-aldehyd (I) resultiert. Auf ähnliche Weise lassen sich auch im Benzolkern substituierte Wieland-Gumlich-aldehyde, z. B. die 10-Chloroverbindung, gewinnen. Ausgangsmaterial hierfür ist das 10-Chlorostrychnin, das durch Chlorierung von II mittels Chlor in konz. Salzsäure nach Leuchs und Steinborn (4) neben 10,15-Dichloro- und 10,15,19-Trichlorostrychnin entsteht (3).

I läßt sich am $N_{(b)}$ leicht alkylieren (Methylieren, Allylieren). Beim Erwärmen mit Essigsäure/ Natriumacetat oder Pivalinsäure reagieren dann 2 Mol. des quartären Wieland-Gumlich-aldehyds derart miteinander, daß die beiden potentiell aldehydischen Zentren 17 und 17' und die zwei nukleophilen Zentren $N_{(a')}$ und $N_{(a)}$ unter Austritt von 2 Mol. Wasser und Öffnung der Ätherringe einen zentralen Bisazacyclooctadien-Ring ausbilden. Aus I-Methylat resultiert des C-Toxiferin (VI), aus dem I-Allylat das (natürlich nicht vorkommende) Alloferin (VII). Diese doppelte Enamin-Bildungsreaktion ist reversibel.

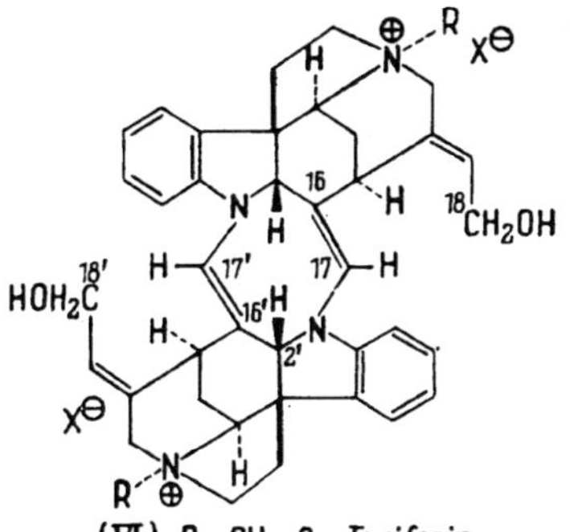

(VI) R=CH_3, *C-Toxiferin*
(VII) R=CH_2–CH=CH_2, *Alloferin*

Durch oxydative Veränderungen des zentralen Achterringes, wie den Einbau einer Ätherbrücke, von C-17-C-17'-, C-16-C-16'-Überbrückungen und von Hydroxylgruppen entstehen andere, natürlich vorkommende Curarealkaloide. Diese Oxydationen lassen sich in vitro nachahmen.

Hier sei nur noch ein Alkaloid erwähnt, nämlich das sich vom 18,18'-Bisdesoxytoxiferin ableitende C-Calebassin, von dem in der Formel VIII nur der zentrale Teil wiedergegeben ist. Diese bisquartäre Base, $C_{40}H_{48}O_2N_4^{\oplus\oplus}$, läßt sich als Folge der C-17-C-17'-Brücke nicht mehr in zwei C_{20}-Bruchstücke spalten. Über eine komplizierte Reaktionsfolge wird C-Calebassin stattdessen mit heißer Mineralsäure in das leuchtend gelb gefärbte Anhydro-isocalebassin-Säureaddukt IX mit der Formel $C_{40}H_{47}ON_4^{\oplus\oplus\oplus} \cdot 3X \cdot H_2O$ umgewandelt (5, 6). Es gelang, die IX zugrunde liegende freie Base X mit Dimethylsulfat in Alkali als stabilen Methyläther XI abzufangen und von seinem kristallisierten Dijodid eine vollständige Strukturanalyse auszuführen (7). Man erkennt, daß in IX bzw. XI die C_2-Achse des C-Calebassins (VIII) fehlt: Die Spektren, insbesondere die NMR-Spektren, sind dadurch fast unanalysierbar kompliziert. Die Farbe von IX ist auf das Cyanin-System Ar-NH-C= =C-C=N$^{\oplus}$-Ar zurückzuführen.

Über die Biosynthese der Curarealkaloide ist experimentell noch nicht gearbeitet worden. Da der monomere Wieland-Gumlich-aldehyd (I) und sein $N_{(b)}$-Methosalz in südamerikanischen Strychnaceen auftreten, sind sie als unmittelbare Vorläufer der "dimeren" Alkaloide in Betracht zu ziehen.*)

Das erste, der zweiten Gruppe zuzuschreibende Alkaloid, das strukturell aufgeklärt wurde, ist das erstmals im Jahr 1887 isolierte Geissospermin (XII) (9). Der Stoff läßt sich durch wässerige Mineralsäure zur Hauptsache zu dem mit dem Wieland-Gumlich-aldehyd (I) verwandten Geissoschizolin (XIII) und dem Corynanthein ähnlichen Geissoschizin (XIV) hydrolysieren und aus beiden Partnern durch milde Säurekatalyse resynthetisieren. Man beachte, daß XII und XIII an C-16 die umgekehrte

*) Zusammenfassungen über Curare-Alkaloide vgl. (6, 8).

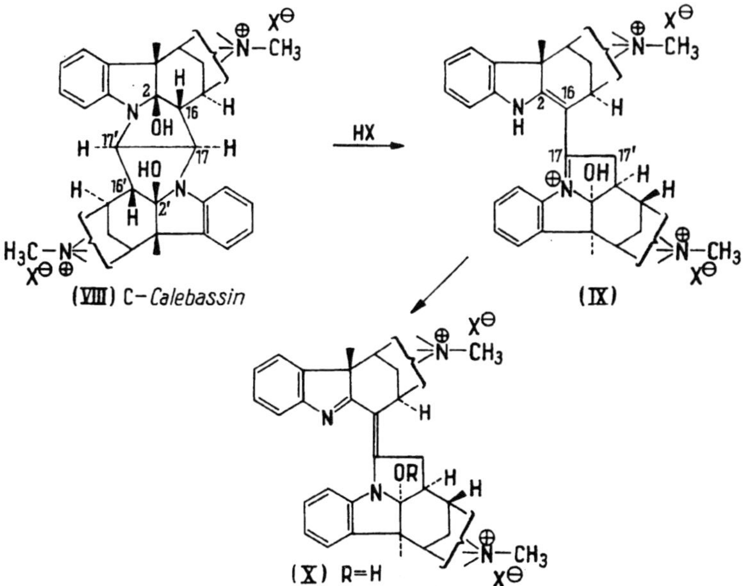

(VIII) C-Calebassin (IX) (X) R=H (XI) R=CH₃

Konfiguration besitzen wie Wieland-Gumlich-aldehyd (I).

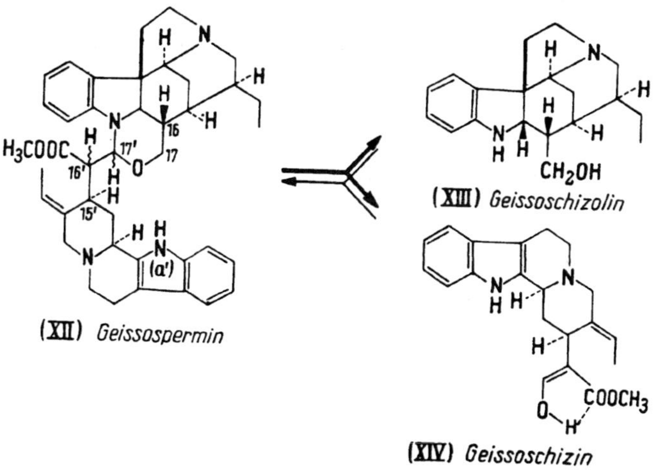

(XII) *Geissospermin* (XIII) *Geissoschizolin* (XIV) *Geissoschizin*

Für Geissospermin gibt es noch eine alternative Formel mit dem Strukturelement C-17-O-C-17'-$-N_{(a')}$. Diese wird u. a. durch das Massenspektrum ausgeschlossen, das durch die Fragmente m/e 381 und 251 gekennzeichnet ist, die durch Spaltung der allylisch aktivierten Bindung C-15'-C-16' resultieren und zusammen das Molekulargewicht von XII ergeben (10).

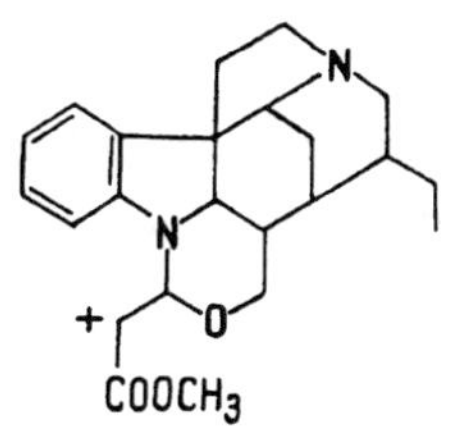

m/e 381

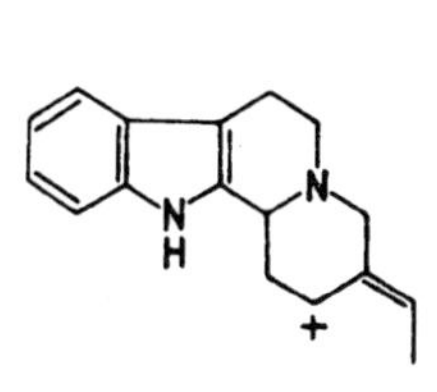

m/e 251

Die Vertreter der zweiten Gruppe sind so mannigfaltig und zahlreich, daß hier nur noch einige Typen diskutiert werden können. Um die große Variabilität aufzuzeigen, seien zunächst Bisindole aufgeführt, bei denen die eine Komponente stets Pleiocarpamin (XV) repräsentiert. Es sind dies die Apocynacea-Alkaloide Villalstonin (XVI) (11), Pycnanthin (XVII) (12) und Pycnanthinin (XVIII) (13). Die Stellen 7 und 2 von XV sind das nucleophile bzw. nach Reaktion der Stelle 7 das elektrophile Reaktionszentrum für den zweiten Partner.

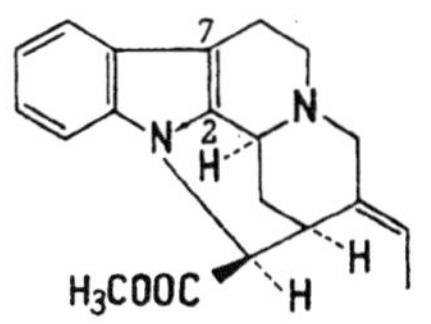

(XV) *Pleiocarpamin*

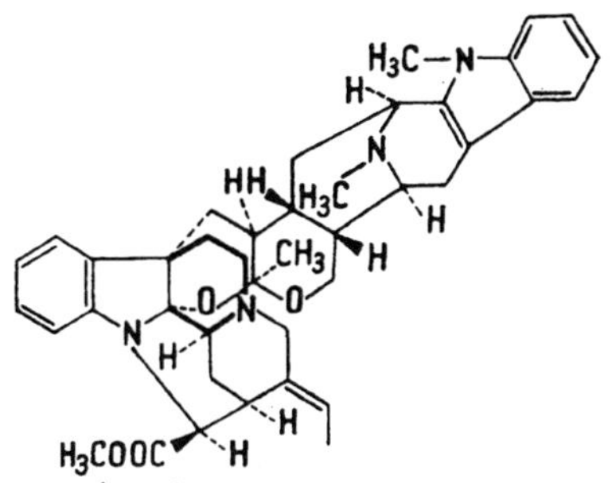

(XVI) *Villalstonin*

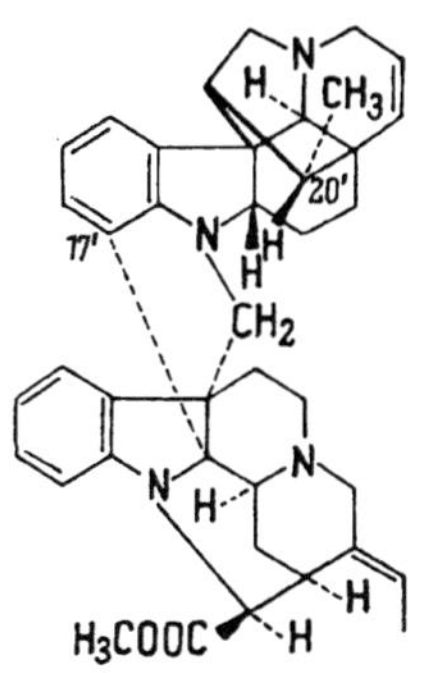

(XVII) *Pycnanthin*

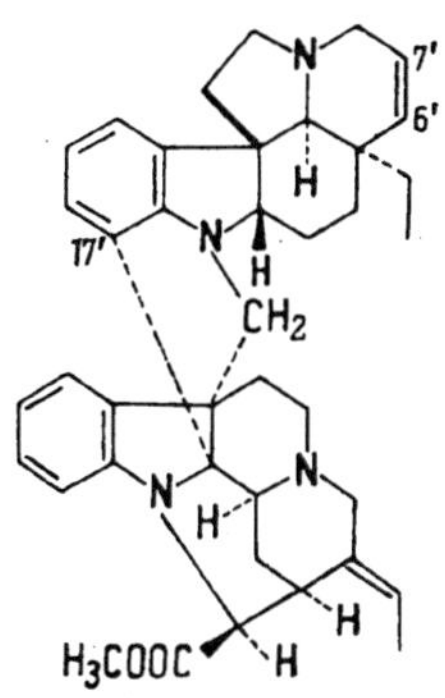

(XVIII) *Pycnanthinin*

Säurekatalysierte Hydrolyse von Pycnanthin (XVII) gab Formaldehyd (aus der Methylenbrücke), Pleiocarpamin (XV) und den oberen Teil in Form von 20-epi-Tuboxenin (= Isotuboxenin) und Tuboxenin. Ähnlich verhält sich Pycnanthinin (XVIII) bei der Hydrolyse: neben Formaldehyd und (XV) wird 6,7-Dehydroaspidospermidin gebildet. Die Tatsache, daß bei XVII und XVIII Pleiocarpamin (XV) mit der Indoldoppelbildung mit dem $N_{(a')}$ und einer aromatischen Position des Indolinteiles der oberen Hälfte reagiert, macht wahrscheinlich, daß die Biosynthese von XVII und XVIII aus den "fertigen" monomeren Alkaloiden etwa gemäß dem umseitigen (oberes) Schema erfolgt.

Der letzte Schritt dieser Mannich-artigen Kondensation ähnelt demjenigen, der vermutlich bei der Biogenese der Alkaloide Pleiomutin (XIX) (14) und Umbellamin (XX) (15) aus (-)-Eburnamin (XXI) und den Indolinen Pleiocarpinin (XXII) bzw. - dem noch unbekannten - 11-Hydroxypseudoakuammigin (XXIII) eine Rolle spielt. Die Biosynthese von XIX ließ sich in vitro durch Kondensation von XXI und XXII mit verd. Schwefelsäure nachahmen (14). Die beiden Alkaloide lassen sich mit heißer Mineralsäure (Retro-Biogenese) "spalten". Bei beiden wird der Eburnamin-Teil dabei zerstört; von Umbellamin (XX) überlebt auch der Pseudoakuammigin-Teil die Hydrolyse nicht; durch Thermo-

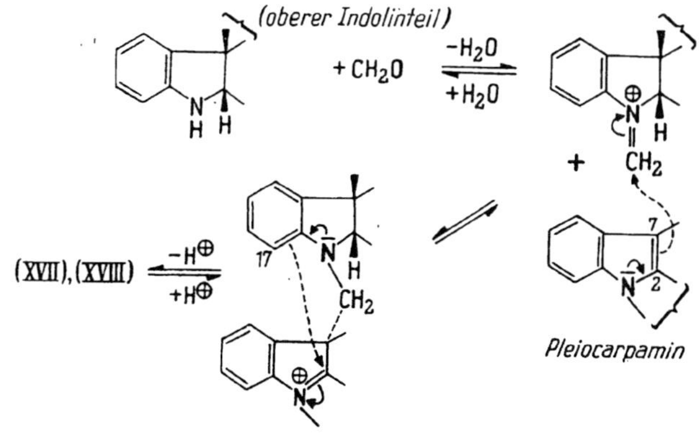

lyse liefert dieses Alkaloid hingegen den Eburnamin-Teil in Form von Eburnamenin, dem Wasserabspaltungsprodukt von XXI.

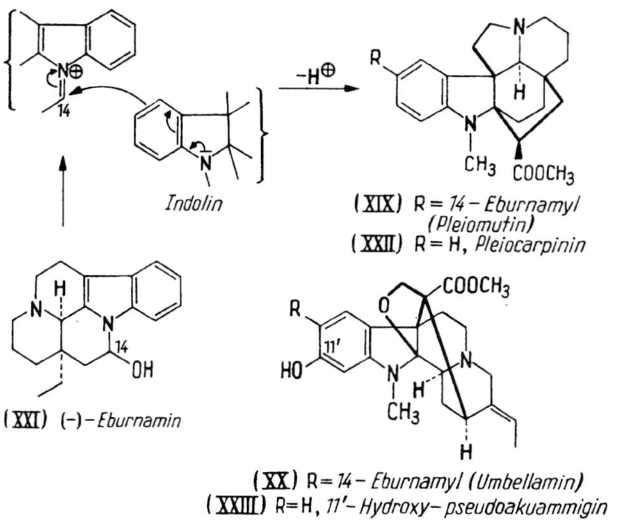

Ein Alkaloid aus zwei typgleichen Hälften repräsentiert das Macralstonin (XXIV) (16). Die eine Hälfte ist das Alstophyllin (XXV), die andere das Macrolin, das sich von XXV durch das Fehlen der aromatischen Methoxylgruppe und geringe Veränderungen im Ring E unterscheidet: Bezeichnenderweise ist das durch die Methoxylgruppe des Alstophyllins (XXV) aktivierte Zentrum 10' mit dem elektrophilen Zentrum 21 des Macrolinteiles verknüpft und nicht umgekehrt.

Reduktive Veränderungen der relevanten Verknüpfungsstellen im biogenetischen Vorläufer für die Bisindolalkaloide vom Vobtusin (XXVI) (17, 18) (und Callichilin-Typ (19)) sind der Grund, weshalb sich diese Alkaloide nicht mehr "spalten" lassen. Chemische Transformation verbunden mit extensivster Auswertung der Massenspektren ermöglichten es für Vobtusin, die Arbeitsformel XXVI vorzuschlagen (18). Zweifellos ist man dabei an die Grenze der mit chemischen und den üblichen spektroskopischen Methoden aufklärbaren Strukturen angelangt. Die Formel für Vobtusin harrt noch der röntgenstrukturanalytischen Bestätigung. Mit dem Vobtusin sind eine Reihe weiterer Alkaloide ver-

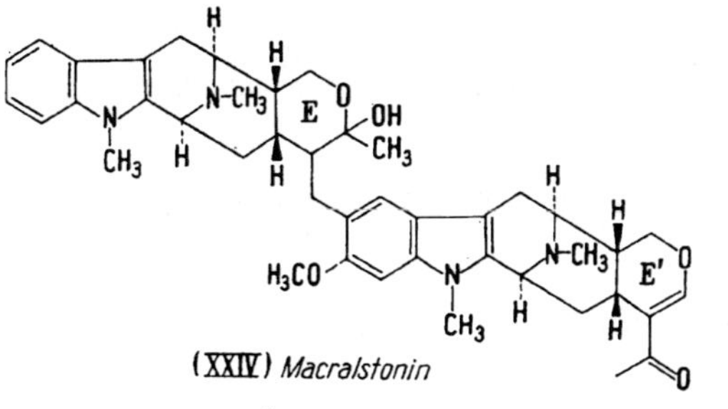

(XXIV) *Macralstonin*

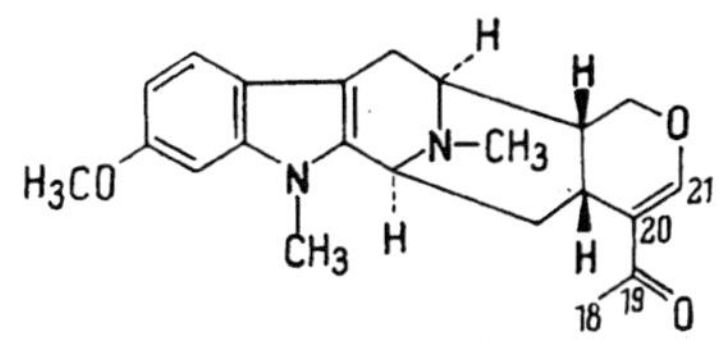

(XXV) *Alstophyllin*

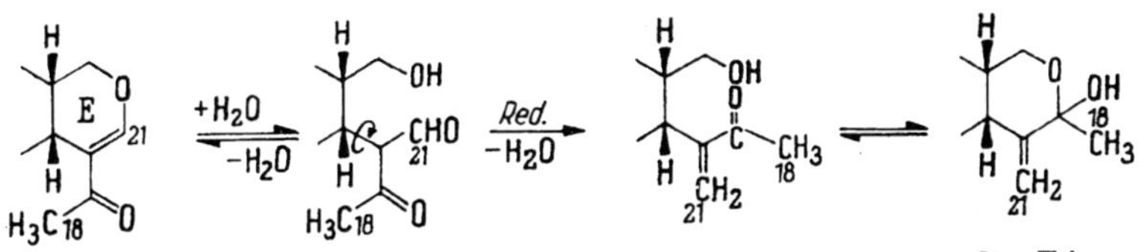

Ring E *im Alstophyllin bzw.* E' *im Macralstonin* — *Ring* E *im Macrolin* — *Ring* E *im Macralstonin*

wandt, wie Desoxy-vobtusin, Vobtusinlacton, Desoxyvobtusinlacton (20), ein Desoxy-epoxy-vobtusin (21) u. a. m.

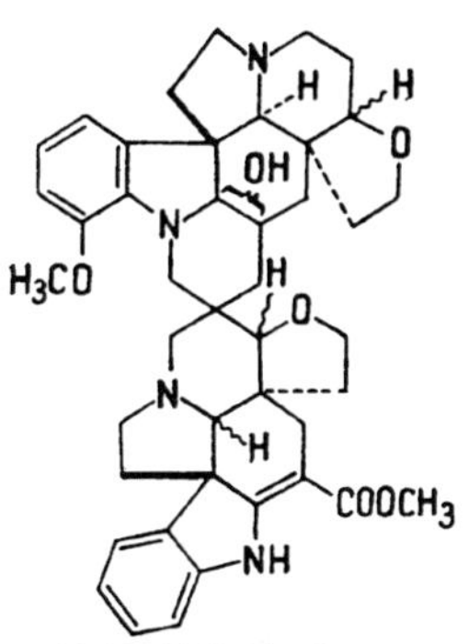

(XXVI) *Vobtusin*

Die in Zürich ausgeführten Arbeiten entstammen der gemeinschaftlichen Alkaloid-Arbeitsgruppe M. Hesse/H. Schmid. Wir danken unseren Mitarbeitern sehr für ihren Einsatz und dem Schweizerischen Nationalfonds für die gewährte Unterstützung.

Literatur

(1) Hesse, M., Indolalkaloide in Tabellen, Berlin, Göttingen, Heidelberg: Springer-Verlag 1964. Ergänzungswerk 1968

(2) Wieland, H. u. W. Gumlich, Liebigs Ann. Chem. 464, 191 (1932): Wieland, H. u. K. Kaziro, Liebigs Ann. Chem. 506, 60 (1933)

(3) Hymon, J.R., H. Schmid, P. Karrer, A. Boller, H. Els, P. Fahrni u. A. Fürst, Helv. chim. Acta, in Vorbereitung

(4) Leuchs, H. u. K. Steinborn, Ber. dtsch. chem. Ges. 71, 1577 (1938)

(5) Bernauer, K., E. Bächli, H. Schmid u. P. Karrer, Angew. Chem. 69, 59 (1957); Bernauer, K., H. Schmid u. P. Karrer, Helv. chim. Acta 40, 731 (1957)

(6) Bernauer, K., Fortschr. Chem. org. Naturstoffe (Wien) 17, 225 (1959)

(7) Gemmell, K.W., J. Robertson, G.A. Sim, K. Bernauer, A. Guggisberg, M. Hesse, H. Schmid u. P. Karrer, Helv. chim. Acta, im Druck 1969

(8) Battersby, A.R. u. H.F. Hodson in R.H.F. Manske "The Alkaloids" VIII, S. 561, Academic Press, New York 1965

(9) Janot, M.-M., Tetrahedron (London) 14, 113 (1961)

(10) Hesse, M., unveröffentlichte Resultate

(11) Hesse, M., H. Hürzeler, C.W. Gemenden, B.S. Joshi, W.I. Taylor u. H. Schmid, Helv. chim. Acta 48, 689 (1965); 49, 1173 (1966)

(12) Gorman, A.A., N.J. Dastoor, M. Hesse, W. von Philipsborn, U. Renner u. H. Schmid, Helv. chim. Acta 52, 33 (1969)

(13) Gorman, A.A., u. H. Schmid, Mh. Chem. 98, 1554 (1967)

(14) Hesse, M., F. Bodmer u. H. Schmid, Helv. chim. Acta 49, 964 (1966); Thomas, D.W., H. Achenbach u. K. Biemann, J. Amer. chem. Soc. 88, 1537 (1966)

(15) Morita, Y., M. Hesse u. H. Schmid, Helv. chim. Acta 52, 89 (1969)

(16) Kishi, T., M. Hesse, W. Vetter, C.W. Gemenden, W.I. Taylor u. H. Schmid, Helv. chim. Acta 49, 946 (1966)

(17) Proisson, J., M. Plat, H. Budzikiewicz, L.J. Durham u. C. Djerassi, Tetrahedron (London) 22, 1075 (1966)

(18) Gorman, A.A., V. Agwada, M. Hesse, U. Renner u. H. Schmid, Helv. chim. Acta 49, 2072 (1966)

(19) Agwada, V., A.A. Gorman, M. Hesse u. H. Schmid, Helv. chim. Acta 50, 1939 (1967)· Plat, M., N. Kunesch, J. Poisson, C. Djerassi u. H. Budzikiewicz, Bull. Soc. chim. France 1967, 2669

(20) Kunesch, J.N., J. Poisson u. B.C. Das, Tetrahedron Letters (London) 1968, 1745

(21) Agwada, V., Dissertation Univ. Zürich 1969

(Aus dem Institut für Biochemie der Pflanzen
der Deutschen Akademie der Wissenschaften zu Berlin, Halle (Saale), DDR)

Biochemie der Steroidalkaloide

Von K. Schreiber

Anläßlich des 3. Hallenser Alkaloidsymposiums vor 4 Jahren wurde in einer zusammenfassenden Übersicht über die bis dahin bekannten, aus Solanaceen-, Apocynaceen-, Buxaceen- und Liliaceen-Gattungen isolierten Steroidalkaloide berichtet. Außerdem wurden die vorliegenden Befunde zur Biosynthese bzw. zum weiteren Stoffwechsel dieser "Pseudoalkaloide" kurz mitgeteilt und die wahrscheinlichen oder zumindest möglichen Biogenesewege zur Diskussion gestellt (1). Abschließend war darauf hingewiesen worden, daß vieles auf diesem Gebiet noch hypothetisch sei und durch künftige Experimente gesichert oder modifiziert werden müsse.

In den seitdem vergangenen Jahren sind auch auf dem Gebiet der Steroidalkaloide wie auf nahezu allen Sektoren der Naturstoffchemie und Biochemie zahlreiche neue Ergebnisse erzielt worden (vgl. u.a. 2-7). Dies gilt auch für das Steroidalkaloid-Vorkommen im Pflanzen- und (allerdings selten) Tierreich. So konnten in einer Vielzahl von Pflanzenarten erstmals Steroidalkaloide nachgewiesen sowie mehrere noch unbekannte Vertreter isoliert und strukturell aufgeklärt werden. Im folgenden seien als Beispiele nur einige wenige solcher Verbindungen angeführt, die zugleich neuartige Strukturtypen repräsentieren.

Wurzeln der im tropischen Südamerika beheimateten und in der dortigen Volksmedizin unter der Bezeichnung 'Jurubeba' genutzten Solanacee Solanum paniculatum L. lieferten das Alkaloidglykosid Jurubin, dessen Struktur als (25 S)-3 β-Amino-5α-furostan-22α.26-diol-0(26)-β-D-glucopyranosid (I) ermittelt wurde (8, 9). Mineralsäure oder enzymatische Hydrolyse dieses Glucosids ergab das mit Jurubidin bezeichnete (25 S)-3 β-Amino-5α.22α0-spirostan(3-Desoxy-3β-amino-neotigogenin, II), bei dem es sich um das erste aus Pflanzenmaterial gewonnene Aminospirostan handelt (9, 10). Das nach Abspaltung der 26-ständigen Glucose primär entstehende Aglykon des Jurubins ist als solches nicht isolierbar, sondern es cyclisiert spontan und stereospezifisch zum Spiroketal der Steroidsapogenine. Solche Ring-E-offenen, im allgemeinen N-freien Furostan-22α.26-diol-0(26)-glykoside sind nach neuesten Untersuchungen offensichtlich in Steroidsaponin-führenden Pflanzen allgemein verbreitet (11-13). Die Auffindung dieser Saponine bestätigte die bereits vor 20 Jahren von Marker und Lopez (14) geäußerte Vermutung, daß es sich bei den Spirostanen um Artefakte handelt, die erst während der Hydrolyse aus sogenannten Prosaponinen mit offener Seitenkette gebildet werden. Jurubin (I) ist kürzlich auch aus Wurzeln von Solanum torvum Sw. gewonnen worden (15).

Gleichfalls neuartige Steroidalkaloidglykoside liegen in den von Janot, Goutarel und Mitarb. (16, 17) aus Holarrhena-Arten (Apocynaceen) isolierten Verbindungen Holacurtin (III) und Mitiphyllin (IV) vor, bei denen an ein stickstofffreies 20-Keto-pregnan bzw. Cardenolid ein Aminozucker, die 4-Desoxy-4-methylamino-D-cymarose, gebunden ist.

Einen weiteren neuartigen, auch biogenetisch interessanten Steroidalkaloidtyp repräsentieren die aus Veratrum album subsp. lobelianum (Bernh.) Suessenguth isolierten (18) Nebenalkaloide Veramin (V) (19, 20), Veralkamin (VI) (21-23) und Veralinin (VII) (24). Bei Veramin handelt es sich um ein 17 β-Methyl-18-nor-16 βH-spirosolan-Derivat, bei Verbindung VI und VII um analoge 16 β-hydroxylierte bzw. 16-unsubstituierte Ring-E-offene Abkömmlinge. In diesen Alkaloiden ist das normale Steroidgerüst durch eine β-seitige 1,2-Umlagerung der angularen 13-Methylgruppe (C-18) an das nachbarständige Kohlenstoffatom 17 modifiziert unter Bildung des in natürlichen Verbindungen bisher noch nicht festgestellten 17 β-Methyl-18-nor-17-iso-cholestan-Kohlenstoffgerüsts. Diese im Sinne einer Sextett-Umlagerung zu formulierende Reaktion führt zu einer α-Stellung der 17-ständigen Seitenkette sowie zugleich zur Ausbildung der Δ^{12}-Doppelbindung. Das vorgeschlagene Biogeneseschema (VIII-XII) findet eine Parallele in der Wagner-Meerwein-Umlagerung von 16α, 17α-Epoxy-20-keto-pregnanen, die ebenfalls Verbindungen mit Δ^{12}- bzw. Δ^{13}-17 β-Methyl-18-nor-17-iso-Struktur liefert.

Neuere strukturanalytische und synthetische Untersuchungen bei Solanocapsin führten zu einer Revision der früher angegebenen Stereochemie von C-16. Danach ist die 16-Sauerstoff-Funktion nicht β-ständig (25, vgl. 1), sondern im Sinne von Formelbild XVI α-ständig angeordnet (26). Damit dürfte Solanocapsin in Übereinstimmung mit dem früher (1) formulierten Schema biogenetisch eng mit Verbindungen vom Typ des Solacongestidins (XIII) und Solafloridins (XIV) korrelliert sein. Beide wurden kürzlich gemeinsam mit 23-Oxo- (XV) und 24-Oxo-solacongestidin aus Solanum congestiflorum Dun. isoliert (27).

Die aus dem kolumbianischen Pfeilgift-Frosch Phyllobates aurotaenia (Phyllobates bicolor) gewonnene Verbindung Batrachotoxinin A, bei der es sich um ein Hydrat (?) des hochtoxischen Batrachotoxins handelt (28, 29), ließ sich gleichfalls strukturell aufklären (30). Hiernach ist Batrachotoxinin A ein Steroidalkaloid sehr ungewöhnlicher Struktur (XVII) mit einer Epoxyäthano-N-methylepimino-Brücke zwischen C-14 und C-18.

Auch auf dem Gebiet der Steroidalkaloid-Biosynthese wurden einige bemerkenswerte Ergebnisse erzielt. So wird sowohl im Ringgerüst mit ^{14}C markiertes Cholesterin als auch 3 β-Hydroxy-pregn-5-en-20-on (XVIII) in Holarrhena-Alkaloide (z.B. XX) eingebaut (31-33), nicht jedoch, wie ursprünglich vermutet (32), auch Progesteron (XIX) (34). Dieses entsteht vielmehr aus XVIII (35), das in Umkehrung der Alkaloidbiosynthese auch aus Holaphyllamin (XX) gebildet werden kann (34). Dieser Befund deutet darauf hin, daß für die Einführung der 3-Aminogruppe in XX eine direkte Substitution der 3 β-Hydroxygruppe von XVIII durch eine NH_2-Gruppe wahrscheinlicher ist als eine Transaminierungsreaktion über ein entsprechendes 3-Keton analog XIX.

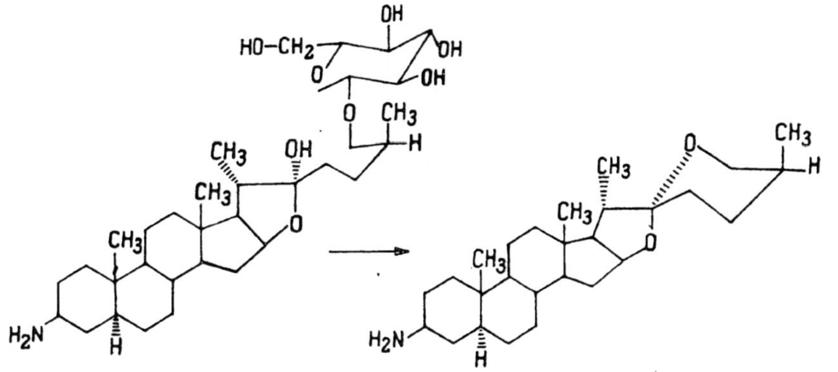

(I) *Jurubin* (II) *Jurubidin*

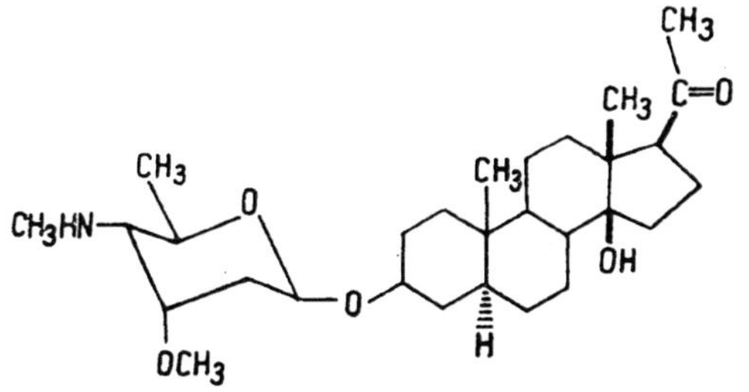

(III) *Holacurtin*

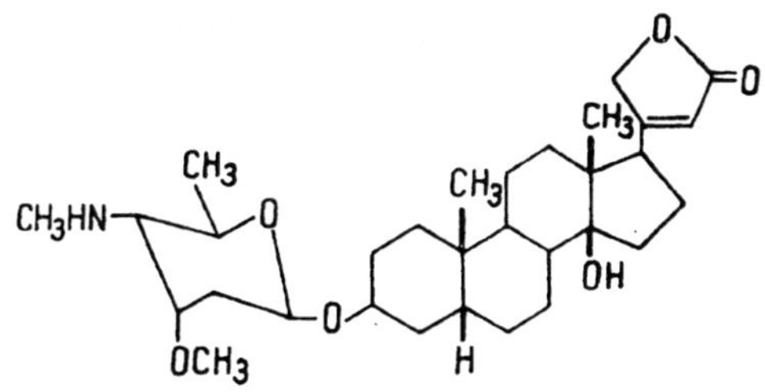

(IV) *Mitiphyllin*

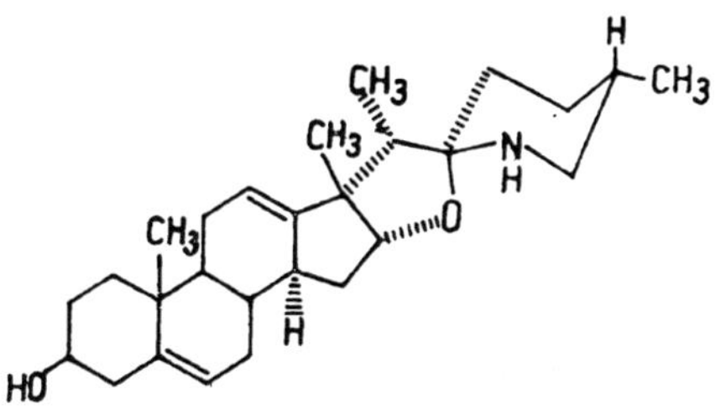

(V) *Veramin*

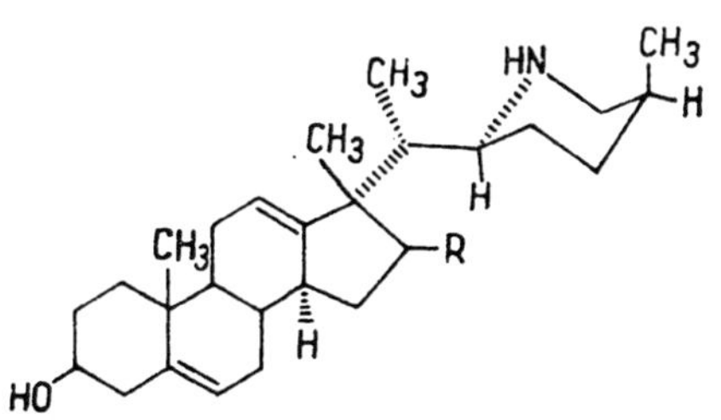

(VI): R=OH *Veralkamin*
(VII): R=H *Veralinin*

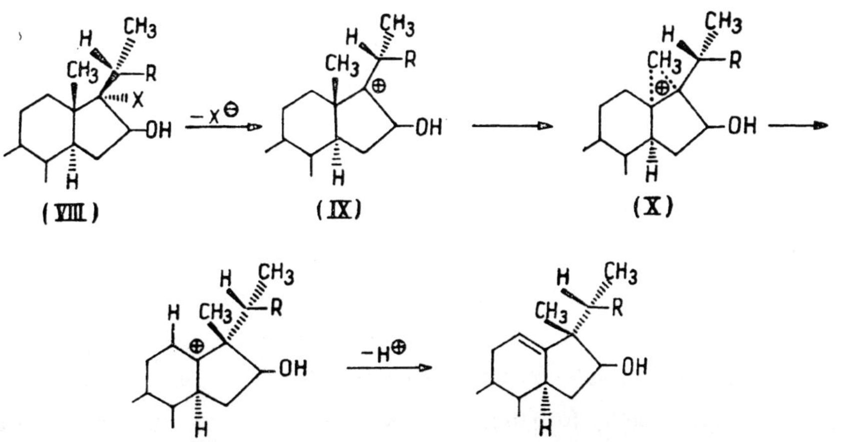

(VIII) (IX) (X)

(XI) (XII)

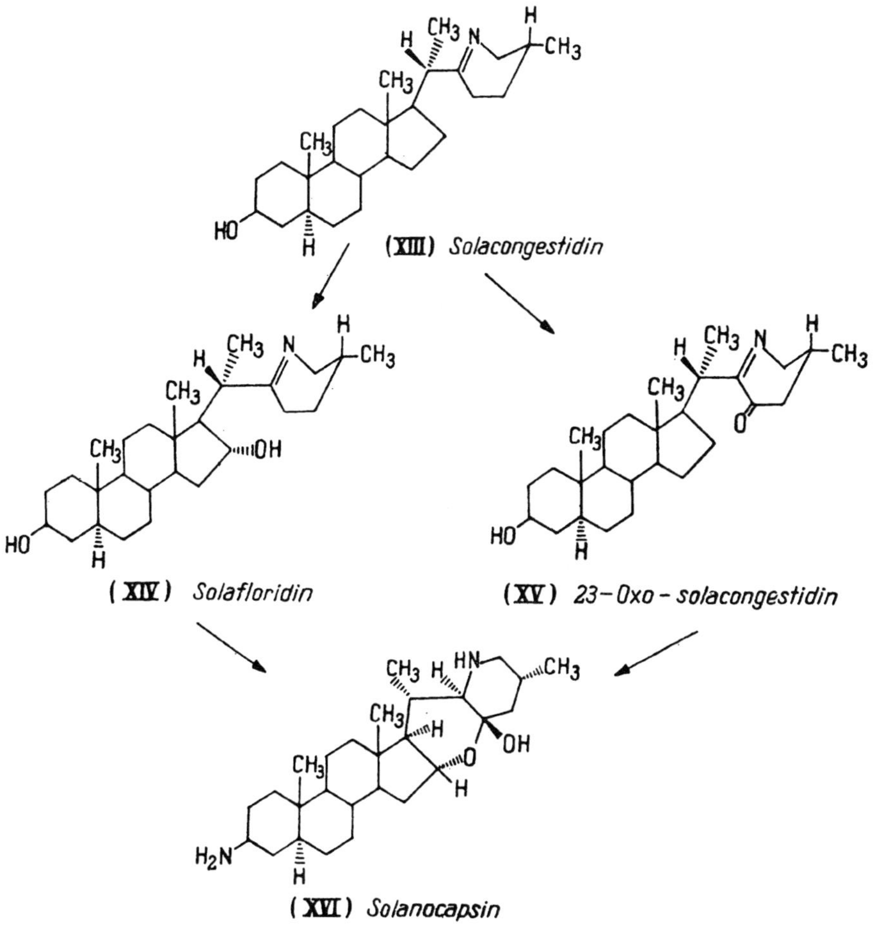

(XIII) *Solacongestidin*

(XIV) *Solafloridin*

(XV) *23-Oxo-solacongestidin*

(XVI) *Solanocapsin*

Cholesterin-(4-^{14}C) (XXI) wird weiterhin eindeutig eingebaut in das Lycopersicon-Alkaloid Tomatidin (XXIII) (36, 37) und das Solanum-Alkaloid Solanidin (XXII) (38) bzw. in die mit diesen chemisch verwandten Steroidsapogenine (z.B. XXIV) (39-41). Jedoch scheinen die stickstofffreien Spirostanole vom Typ XXIV keine Zwischenprodukte der Spirosolanol-Biosynthese (vgl. XXIII) zu sein, denn nach neuesten Untersuchungen von Tschesche und Mitarbeitern (42) wird das Cholesterinderivat Cholest-4-en-3-on-(4-^{14}C) (XXV) zwar in Tigogenin (25-Stereoisomer von XXIV), jedoch nicht in Gitogenin (2α-Hydroxy-tigogenin) oder Tomatidin (XXIII) eingebaut.

Somit dürfte feststehen, daß auch die Steroidalkaloide, und zwar sowohl mit C_{27}-Cholestan- als auch C_{21}-Pregnan-Kohlenstoffgerüst, wie alle weiteren, bereits besser untersuchten Phyto- und Zoosteroide über Cholesterin (XXI) oder eines seiner unmittelbaren Derivate als Zwischenprodukt biosynthetisiert werden. Dieses bis vor kurzem als typisches Zoosterin angesehene Steroid konnte in den letzten Jahren in zahlreichen auch höheren Pflanzen nachgewiesen werden (vgl. 43). Seine Biosynthese in Pflanzen (und hier vermutlich bevorzugt über das Triterpen Cycloartenol) ist in jüngster Vergangenheit mehrfach und systematisch untersucht worden (44, 45, vgl. 43), wobei ein für Pflanze und Tier zum Teil unterschiedlicher stereochemischer Verlauf der Cholesterin-Biosynthese

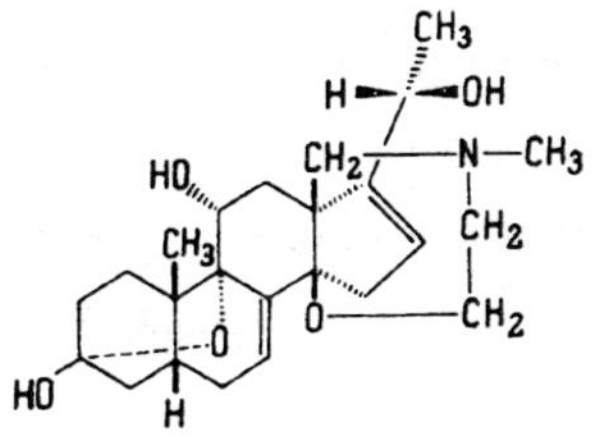

(XVII) *Batrachotoxinin A*

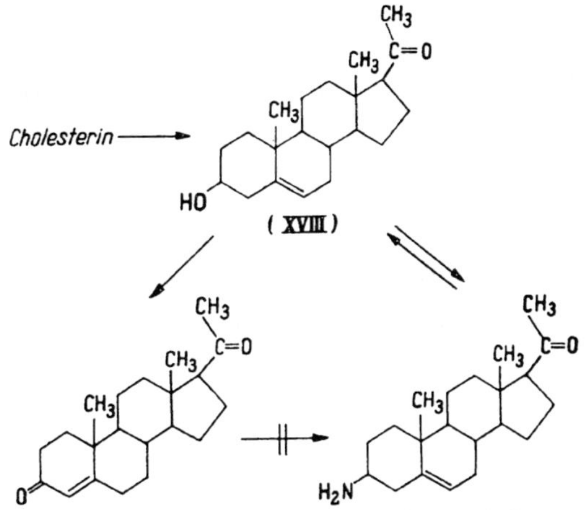

(XVIII)

(XIX) *Progesteron*

(XX) *Holaphyllamin*

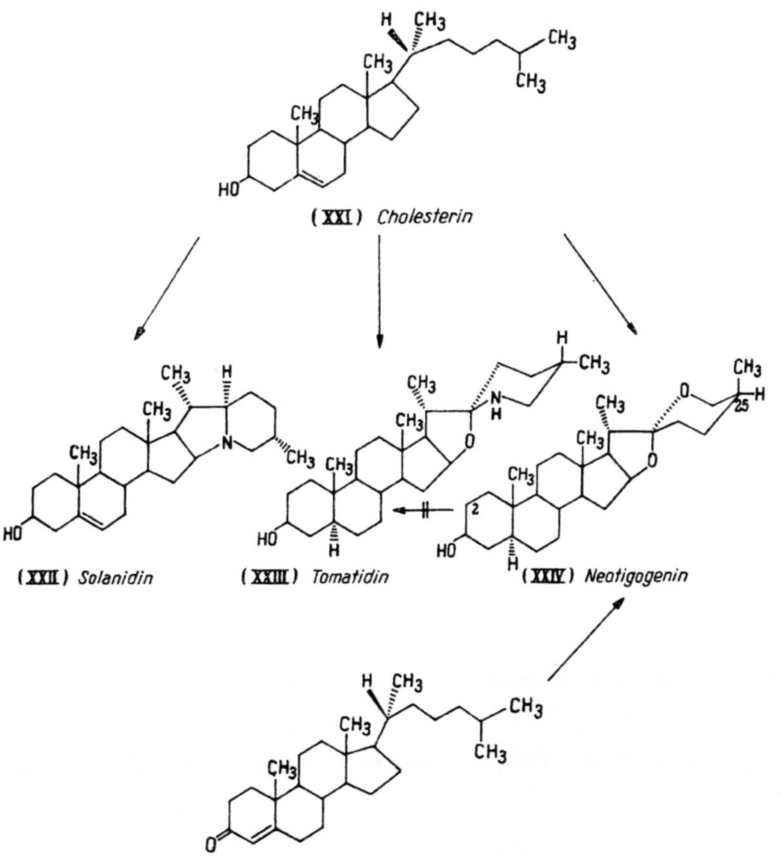

(XXI) *Cholesterin*

(XXII) *Solanidin*

(XXIII) *Tomatidin*

(XXIV) *Neotigogenin*

(XXV) *Cholest-4-en-3-on*

festgestellt wurde und besonderes Interesse verdient (46-48). Für die detaillierte Klärung der Steroidalkaloid- (und Steroidsapogenin-)Biosynthese und weiterer Stoffwechselprozesse sowie deren Regulation auf zellulärer und molekularer Ebene eröffnet die Anwendung von Gewebe- und Zellkulturen höherer Pflanzen neue und in vieler Hinsicht bessere Möglichkeiten (vgl. 49-51).

Literatur

(1) Schreiber, K., Abh. dtsch. Akad. Wiss. Berlin, Kl. Chem., Geol., Biol., Jg. 1966, Nr. 3, 65, Akademie-Verlag, Berlin

(2) Cerný, V., u. F. Sorm, in: R.H.F. Manske, The Alkaloids, Vol. 9, p. 305, New York - London 1967

(3) Habermehl, G., in: R.H.F. Manske, The Alkaloids, Vol. 9, p. 427, New York - London 1967

(4) Schreiber, K., in: R.H.F. Manske, The Alkaloids, Vol. 10, p. 1, New York - London 1968

(5) Kupchan, S.M., u. A.W. By, in: R.H.F. Manske, The Alkaloids, Vol. 10, p. 193, New York - London 1968

(6) Heftmann, E., Lloydia 30, 209 (1967)

(7) Schütte, H.R., in: K. Mothes u. H.R. Schütte, Biosynthese der Alkaloide. Verlag der Wissenschaften, Berlin 1969, S. 616

(8) Schreiber, K., u. H. Ripperger, Tetrahedron Letters (London) 1966, 5997

(9) Ripperger, H., H. Budzikiewicz u. K. Schreiber, Chem. Ber. 100, 1725 (1967)

(10) Schreiber, K., H. Ripperger u. H. Budzikiewicz, Tetrahedron Letters (London) 1965, 3999

(11) Tschesche, R., G. Lüdke u. G. Wulff, Tetrahedron Letters (London) 1967, 2785

(12) Tschesche, R., B.T. Tjoa, G. Wulff u. R.V. Novonka, Tetrahedron Letters (London) 1968, 5141

(13) Kiyosawa, S., M. Hutoh, T. Komori, T. Nohara, I. Hosokawa u. T. Kawasaki, Chem. Pharmac. Bull. (Tokyo) 16, 1162 (1968)

(14) Marker, R.E., u. J. Lopez, J. Amer. chem. Soc. 69, 2389 (1947)

(15) Schreiber, K., u. H. Ripperger, Kulturpflanze 15, 199 (1967)

(16) Janot, M.M., P. Devissaguet, K.-H. Qui, J. Parello, N.G. Bisset u. R. Goutarel, C.R. hebd. Séances Acad. Sci. C 266, 388 (1968)

(17) Janot, M.M., M. Leboeuf, A. Cavé, R. Wijesekera u. R. Goutarel, C.R. hebd. Séances Acad. Sci. C 267, 1050 (1968)

(18) Tomko, J., u. A. Vassová, Chem. Zvesti 18, 266 (1964)

(19) Adam, G., K. Schreiber, J. Tomko, Z. Votický u. A. Vassová, Tetrahedron Letters (London) 1968, 2815

(20) Tomko, J., Z. Votický, A. Vassová, G. Adam u. K. Schreiber, Collect. czech. chem. Commun. 33, 4054 (1968)

(21) Tomko, J., A. Vassová, G. Adam, K. Schreiber u. E. Höhne, Tetrahedron Letters (London) 1967, 3907

(22) Tomko, J., A. Vassová, G. Adam u. K. Schreiber, Tetrahedron (London) 24, 4865 (1968)

(23) Höhne, E., G. Adam, K. Schreiber u. J. Tomko, Tetrahedron (London) 24, 4875 (1968)

(24) Tomko, J., A. Vassová, G. Adam u. K. Schreiber, Tetrahedron (London) 24, 6839 (1968)

(25) Schreiber, K., u. H. Ripperger, Liebigs Ann. Chem. 655, 114 (1962)

(26) Ripperger, H., u. K. Schreiber, Liebigs Ann. Chem. (im Druck)

(27) Sato, Y., Y. Sato, H. Kaneko, E. Bianchi u. H. Kataoka, J. org. Chemistry (im Druck)

(28) Märki, F., u. B. Witkop, Experientia (Basel) 19, 329 (1963)

(29) Daly, J.W., B. Witkop, P. Bommer u. K. Biemann, J. Amer. chem. Soc. 87, 124 (1965)

(30) Tokuyama, T., J. Daly, B. Witkop, I.L. Karle u. J. Karle, J. Amer. chem. Soc. 90, 1917 (1968)

(31) Bennett, R.D., u. E. Heftmann, Arch. Biochem. Biophysics 112, 616 (1965)

(32) Bennett, R.D., u. E. Heftmann, Phytochemistry 4, 873 (1965)

(33) Tschesche, R., u. H. Hulpke, Z. Naturforsch. 23b, 283 (1968)

(34) Bennett, R.D., E. Heftmann u. S.-T. Ko, Phytochemistry 5, 517 (1966)

(35) Bennett, R.D., u. E. Heftmann, Science (Washington) 149, 652 (1965)

(36) Tschesche, R., u. H. Hulpke, Z. Naturforsch. 21b, 893 (1966)

(37) Heftmann, E., E.R. Lieber u. R.D. Bennett, Phytochemistry 6, 225 (1967)

(38) Tschesche, R., u. H. Hulpke, Z. Naturforsch. 22b, 791 (1967)

(39) Bennett, R.D., u. E. Heftmann, Phytochemistry 4, 577 (1965)

(40) Tschesche, R., u. H. Hulpke, Z. Naturforsch. 21b, 494 (1966)

(41) Bennett, R.D., E.R. Lieber u. E. Heftmann, Phytochemistry 6, 837 (1967)

(42) Tschesche, R., H. Hulpke u. R. Fritz, Phytochemistry 7, 2021 (1968)

(43) Goad, L.J., in: J.B. Pridham, Terpenoids in Plants. S. 159, London - New York 1967

(44) Rees, H.H., L.J. Goad u. T.W. Goodwin, Tetrahedron Letters (London) 1968, 723

(45) Rees, H.H., L.J. Goad u. T.W. Goodwin, Biochem. J. 107, 417 (1968)

(46) Smith, A.R.H., L.J. Goad u. T.W. Goodwin, Chem. Commun. 1968, 926, 1259

(47) Gibbons, G.F., L.J. Goad u. T.W. Goodwin, Chem. Commun. 1968, 1458

(48) Caspi, E., u. P.J. Ramm, Tetrahedron Letters (London) 1969, 181

(49) Heble, M.R., S. Narayanaswami u. M.S. Chadha, Naturwissenschaften 55, 350 (1968)

(50) Heble, M.R., S. Narayanaswami u. M.S. Chadha, Science (Washington) 161, 1145 (1968)

(51) Kaul, B., u. E.J. Staba, Lloydia 31, 171 (1968)

(From the Chemistry and Life Sciences Laboratory,
Research Triangle Institute, U.S.A.)

Alkaloids with Anti-Tumor Activity

By M. E. Wall

Introduction

The use of plants as cancer remedies goes back to the ancient Egyptians (1). Hartwell (2) has listed over 3.000 plants which in folk-lore over the centuries, and some even up to the present time, are used in all parts of the world. Needless to say, most of these remedies have been useless. During the last 10-15 years, however, a large effort has been devoted to a scientific study of higher plants as sources of new anti-tumor agents, supported in the U.S.A. and in many other parts of the world by the Cancer Chemotherapy National Service Center, National Cancer Institute, National Institutes of Health. In addition some pharmaceutical companies, notably Eli Lilly and Company in the U.S.A., have developed and marketed a plant antitumor agent.

These more recent studies have been guided heavily by bioassay. Plant extracts are initially checked for their ability to inhibit the growth of a tissue cell culture or an experimental solid tumor in mice or rats, or for their ability to increase the life span of mice treated with various leukemias. If activity is found, fractionation proceeds using all the modern techniques available, including combinations of Craig Counter-Current Distribution, adsorption and partition chromatography, thin-layer chromatography, ion exchange chromatography and various molecular sieves. In successful applications fractionation of the activity occurs and anti-tumor activity is observed with lower and lower doses of the purified extract, culminating in the isolation of a crystalline product. Bioassay is a very powerful detection tool, and in the writer's experience has permitted the isolation of many compounds which are found in only trace amounts (0.001-0.01 % dry weight) in plants. In particular alkaloids have been isolated which have actual use or promising possibilities in cancer chemotherapy.

This paper will deal only with a limited number of alkaloids from higher plants which have shown promising results in experimental tumor systems or in the clinic. A brief review of the work of other research groups will be presented along with recent work from the author's laboratory. The author wishes in particular to thank Dr. J.L. Hartwell for permission to reproduce data from a recent review of this subject (3).

Vinblastine (VLB) and Vincristine (VCR)

Vinblastine (I) and Vincristine (II) are the first two alkaloids to be isolated which are actively used in the clinic. The original discovery by Noble and colleagues owes much to serendipity (4) as the alkaloids were isolated during a study of extracts of the Periwinkle [Catharanthus roseus (Vinca rosea)] in the hope of finding an anti-diabetic agent, but finding instead that marked depression of white blood cells occurred.

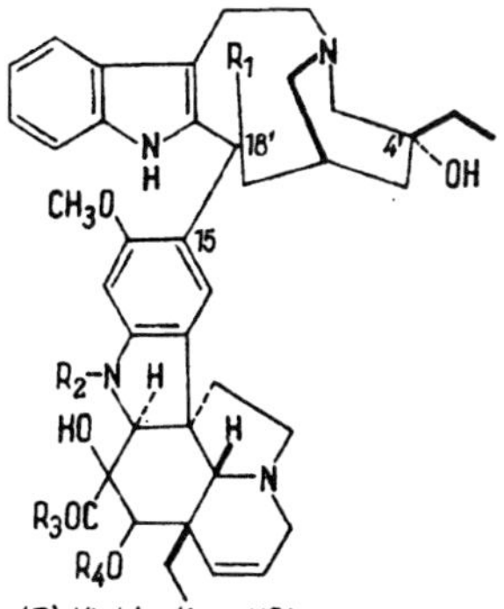

(I) *Vinblastine, VBL*
R_1=$COOCH_3$; R_2=CH_3; R_3=OCH_3; R_4=$COCH_3$
(II) *Vincristine, VCR*
R_1=$COOCH_3$; R_2=CHO; R_3=OCH_3; R_4=$COCH_3$

The chemistry of these dimeric indole-indoline alkaloids have been reviewed in detail by Svoboda (5) and Neuss, Gorman et al. (6, 7, 8). VCR and VLB were initially active in life prolongation using the P-1534 leukemia system. VCR and VLB are both in active clinical use; VLB is particularly useful in Hodgkin's Disease and VCR in acute leukemia (children). VCR and VLB are often used in conjunction with prednisone, methotrexate and 6-mercaptopurine. Excellent reviews have been recently presented by Bernard (9) and Frei (10).

Tylocrebrine and Related Alkaloids from Tylophora crebiflora

The chemistry of these phenanthroindolizene alkaloids has been reviewed by Govindachari (11a), the pioneer worker in this field. Rao and co-workers (11b, 11c) at Charles Pfizer and Co. have recently isolated a number of new alkaloids as well as some previously known compounds from Tylophora crebiflora (12a). The basic structure (III) is shown below and in structural variants as shown in Table 1 (reference 3).

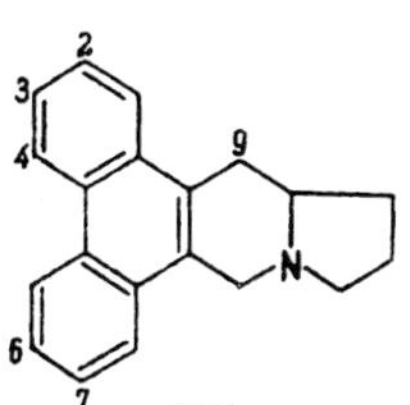

(III)

Table 1: Structure of Alkaloids from Tylophora crebriflora

Alkaloid	Substituents at C-atom					
	2	3	4	6	7	9
Tylocrebrine	-	OCH_3	OCH_3	OCH_3	OCH_3	-
Tylophorine	OCH_3	OCH_3	-	OCH_3	OCH_3	-
Tylophorinine	-	OCH_3	-	OCH_3	OCH_3	OH
Compound A	-	OCH_3	OCH_3	OCH_3	OCH_3	OH
Compound B	-	OCH_3	OH	OCH_3	OCH_3	-
Compound C	-	OCH_3	OH	OCH_3	OCH_3	OH
Compound D	OCH_3	OCH_3	OCH_3	OCH_3	OCH_3	OH
Compound E	OCH_3	OCH_3	OCH_3	OCH_3	OCH_3	-
Compound F	Structure unknown					

Gellert and Rudzats (12b) have also shown that tylocrebrine is active against L-1210 leukemia. Almost all of the compounds show moderate to good activity in various solid tumor and leukemia systems (3). Tylocrebrine which occurs in largest quantity in T. crebriflora shows, among orther activities, inhibition of Sarcoma 180 and 50-70 % increase in life span with the L-1210 and P-388 leukemia systems. The entire group of alkaloids is noteworthy in that all the analogues also show anti-neoplastic activity. Tylocrebrine unfortunately demonstrated irreversible central nervous system toxicity and is not under clinical testing at present.

Ellipticine and Related Compounds

Ellipticine (6-H-pyrido [4,3b] -carbazole, 5, 11-dimethyl) was first isolated from Ochrasia elliptica (13). The chemistry of ellipticine (IV) has been reviewed by Gilbert (14). Ellipticine has been synthesized by Dalton et al. (15) and many analogues have also been made. As a consequence, there is a large body of structure-function activity data available (3). Ellipticine has excellent life prolongation activity in the L-1210 and P-388 leukemia systems. Variations in the basic theme including 5-methyl, 6-methyl, 1,5-dimethyl, 5,11-dimethyl (ellipticine), 9-methoxy-5,11-dimethyl, are found in which activity is not destroyed. However substitution of methyl at 3 or 9 and of bromine at 9 destroy activity. Ellipticine is a candidate for advanced pharmacological studies prior to possible clinical testing.

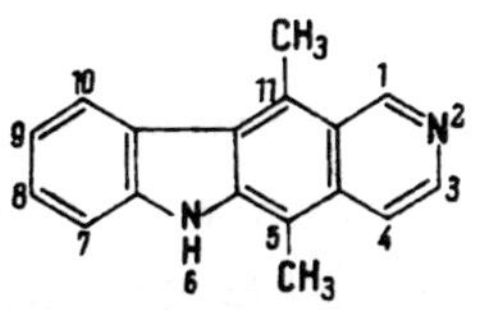

(IV) Ellipticine

Bisbenzylisoquinoline Alkaloids

The chemistry of this class of alkaloids has recently been reviewed by Curcumelli-Rodoslamo and Kulka (16). A large number of alkaloids in this group have been tested (3). Significant activity in the Walker Intramuscular Carcinosarcoma (WM-256) has been observed with several alkaloids isolated by Kupchan et al. (17, 18). All are inactive in L-1210 leukemia. Typical representatives of this series are thalicarpine (V), D-, L-, and DL-tetrandine (VI). Thalicarpine and D-tetrandine have been selected for preclinical pharmacological study.

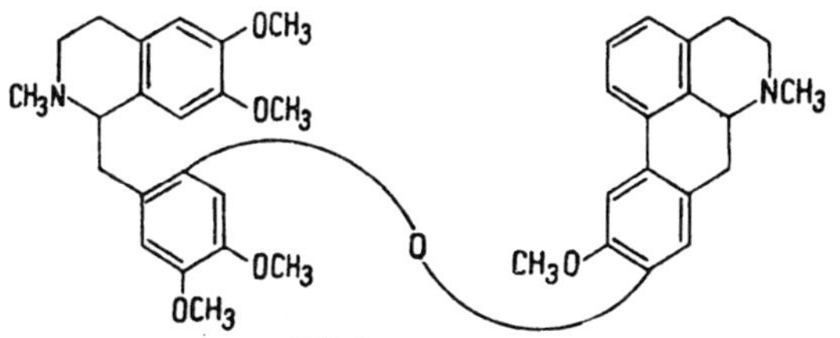

(V) *Thalicarpine*

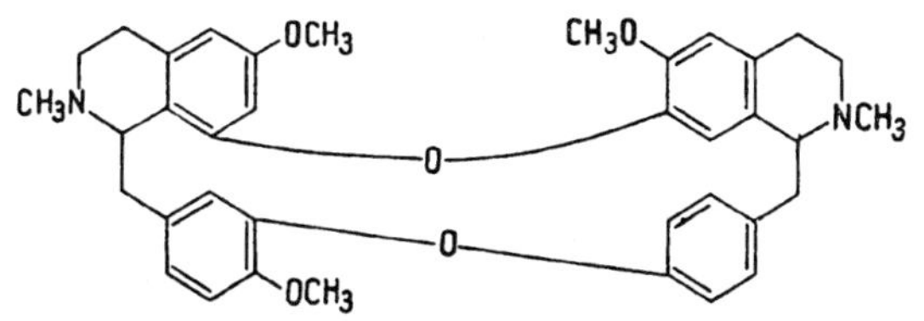

(VI) *DL-Tetrandine*

Acronycine

Acronycine (VII) was isolated by Svoboda from Acronychia banen (baueri australiana) (19, 20). Many other related alkaloids and a variety of acridines and acridones were compared with (VII) for anti-tumor activity (20). Although acronycine displayed a broad spectrum of anti-tumor activity in a variety of experimental neoplasms and leukemias (19, 20), all the other related compounds were inactive.

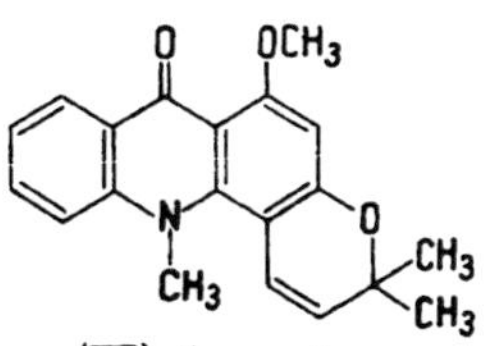

(VII) *Acronycine*

Camptothecin and Related Alkaloids

Camptothecin was isolated from the bark and wood of Camptotheca acuminata, a tree which is a native of China and belongs to the family Nyssaceae. No other species are found in this family. First chemical studies on this species was conducted by Wall et al. about 15 years ago (21). The plant gave no significant test for saponins, alkaloids and flavonoids. Some years later, several thousand crude alcoholic extracts from the author's laboratory (then at Eastern Utilization Research and Development Division,

ARS, USDA) were tested by CCNSC for anti-tumor activity. C. acuminata gave excellent inhibition in the CA-755 tumor and hence was selected for further fractionation. Schemes 1 and 2 show the method of initial isolation and demonstrate (if such a demonstration is necessary) the powerful tool which a good bioassay system can be for the isolation of biologically active compounds. It should be noted, in passing, that normal methods for alkaloid isolation would be useless for camptothecin as it is not a basic compound. It should be noted, also, that C. acuminata is one of the few plants in the CCNSC screen which gives strong activity in the L-1210 system with crude extracts. From Scheme 1 it can be seen that camptothecin is not extracted from the plant by hexane, but can be removed with hot 95 % ethanol and on partition of the residue from evaporation of ethanol with chloro-

<u>Scheme 1:</u> Preliminary Concentration of Camptothecin (Camptotheca acuminata B654996)

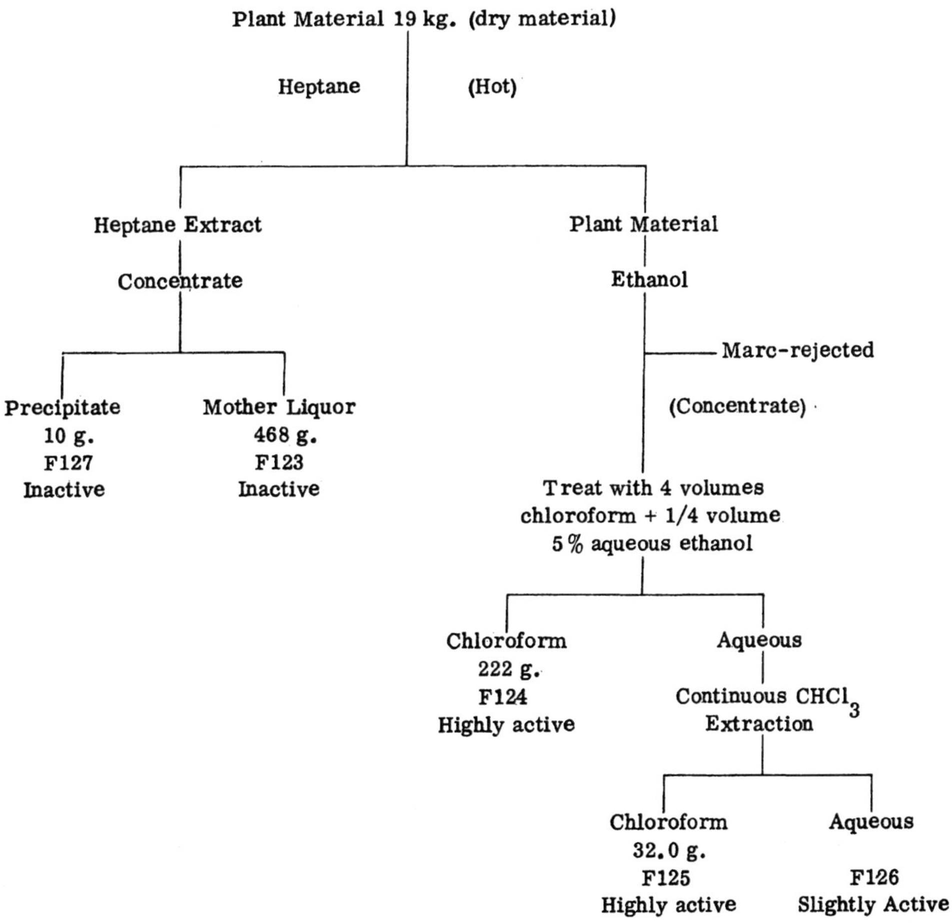

F123-T/C < 125 from 200-25 mg/kg
F127-T/C < 125 from 200-25 mg/kg
F124-T/C = 209 at 50; 163 at 25 mg/kg
F125-T/C = 214 at 50; 151 at 25 mg/kg
F125-T/C = 128 at 200; < 125 at 100-25 mg/kg

Scheme 2: Purification and Isolation of Camptothecin

182 grams chloroform extract (cf. Scheme 1) were subjected to an 11 stage preparative Craig Counter-Current Distribution (System $CCl_4/CHCl_3/MeOH/H_2O$; Ratio 7:3:8:2 respectively, mutually saturated, approximately 900 ml. in upper and lower phase respectively.

B654996

NSC No.	Tube No.	Wt. Fraction (Grams)	Dose (mg./kg.)**	T/C	9KB
F098	-*	-	125	153	< 1.0
F099	1	21.5	250	179	5.3
F0100	2	8.7	62.5	201	0.8
F0101	3	7.2	62.5	201	0.5
F0102	4	6.9	31.2	206	0.09
F0103	5	7.6	31.2	156	0.3
F0104	6	7.2	31.2	198	0.5
F0105	7	9.3	31.2	160	< 1.0
F0106	8	11.5	250	172	0.5
F0107	9	16.1	250	148	4.3
F0108	10	19.2	250	96	33.0
F0109	11	62.7	250	100	> 100.0
K128 (Camptothecin)	-	-	0.5	163	0.7

* Starting material.

** Dose at which maximum T/C in LE-1210 was observed.

form-water goes into the chloroform layer. In Scheme 2 it can be noted that a simple large scale Craig Counter-Current Distribution also increased activity. The most active fractions, tubes 3-6, are active at lower doses than most of the other tubes, and can be compared with the activity of pure camptothecin which is also presented in Scheme 2. The most active fractions thus contained between 1-2 % camptothecin. On combining all these fractions and concentrating the solvent, a precipitate was noted which on testing contained all the activity of the sample. Further purification was obtained by chromatography on silica gel and crystallization. A preliminary communication on the isolation of camptothecin has been made (22). The more important properties of camptothecin are listed in Table 2. The structure of the compound rests largely on UV, IR, NMR, mass spectra, and finally x-ray analysis with a limited amount of chemical data. The evidence on which the structure is based is as follows:

Table 2: Properties of Camptothecin

Light yellow needles, m.p. 264-267° (dec.)

Intense blue fluorescence under UV

$[\alpha]_D^{25}$ + 31.3°

M^+ at m/e 348.1117

Calcd. for $C_{20}H_{16}N_2O_4$ - 348.1111

Qualitative Reactions

1. Negative phenol ($FeCl_3$)
2. Negative Indole Tests
3. Negative Dragendorff and Mayer Tests
4. Gave no crystalline salts with a variety of acids
5. Could not be methylated with diazomethane or dimethyl sulfate under a variety of conditions
6. No reactions with bicarbonate or carbonate but is quantitatively converted to the sodium salt with NaOH at room temperature. On acidification, Na salt regenerates camptothecin

1. The molecular weight of camptothecin is 348 (high resolution mass spectrometry) and this is in best accord with formula $C_{20}H_{16}N_2O_4$ (verified by C, H, N analysis). The compound makes a monoacetate and monochloroacetate with mass spectral composition and CHN analyses in accord with the above formulation. From this data it would seem that camptothecin is a monohydroxy compound which from its molecular formula must be highly unsaturated and made up of a number of rings. The ultraviolet spectrum (Figure 1) and the IR spectrum (Figure 2) and the nmr spectrum of the chloroacetate (used for favorable solubility reasons; camptothecin is almost insoluble in all solvents) all confirm the highly unsaturated, aromatic nature of camptothecin. The infrared spectrum of camptothecin shows a broad band at 1740-1760 cm^{-1} (due either to a lactone or a phenolic ester, and a strong band centered at 1665 cm^{-1} which would initially be due either to presence of a $-C=C-\overset{O}{\overset{\|}{C}}-C-$ moiety or to an amide group. The presence of the lactone was unequivocally proven by treating camptothecin with one equivalent of base. Under these conditions the sodium salt is formed with correct analysis and infrared showing disappearance of 1740-1760 cm^{-1} band and presence of a strong band due to carboxylate anion at 1550 cm^{-1}. The presence of an unsaturated carbonyl group could be ruled out because the peak at 1665 cm^{-1} is not affected by borohydride reduction. Thus all four oxygen functions can be accounted for by 1-<u>OH</u>, 1-N-$\overset{O}{\overset{\|}{\underline{C}}}$-, and 1- $\left(\begin{smallmatrix}C=O\\O\end{smallmatrix}\right.$ function. One nitrogen is also accounted in the -<u>N</u>-$\overset{O}{\overset{\|}{C}}$ linkage, and the other could by biogenetic analogy be placed in one of the aromatic rings. NMR analysis also clearly showed that one ethyl group was substituted on a non-aromatic carbon as shown by the presence of a multiplet (2H) at 2.26 δ ($\underline{CH_2}CH_3$) and a triplet (3H) at 1.08 δ ($CH_2\underline{CH_3}$).

At this stage we joined forces with Drs. Sim and McPhail (then at the University of Illinois and determined the structure of camptothecin iodoacetate (prepared from the chloroacetate). The struc-

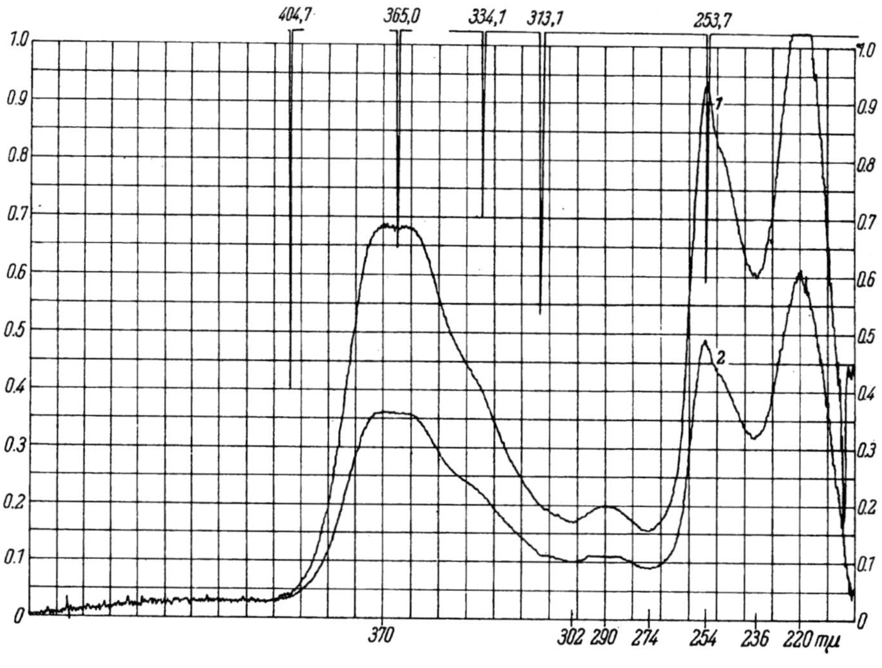

Figure 1. Ultraviolet Spectra of Camptothecin in 95 % Ethanol

Curve 1 0.0112 mg/ml in 95 % Ethanol vs. 95 % Ethanol

Curve 2 0.0056 mg/ml in 95 % Ethanol vs. 95 % Ethanol

λ_{max}		λ_{min}	
mμ	A	mμ	A
220	0.60	236	0.30
254	0.47	274	0.07
290	0.08	302	0.07
370	0.32		

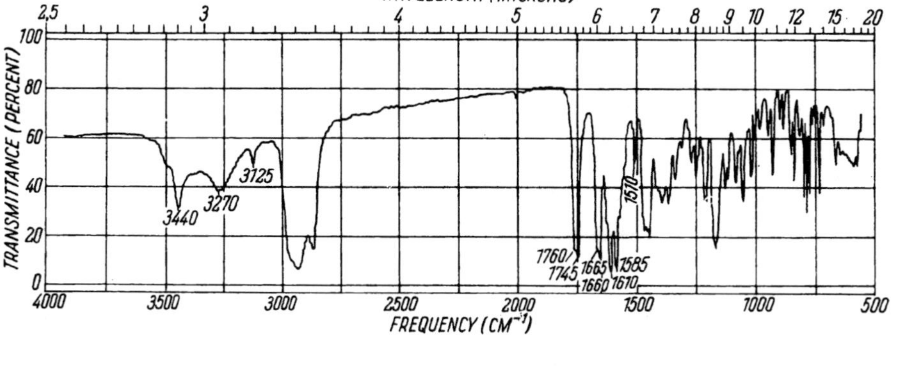

Figure 2. Infrared Spectrum of Camptothecin (Nujol Mull)

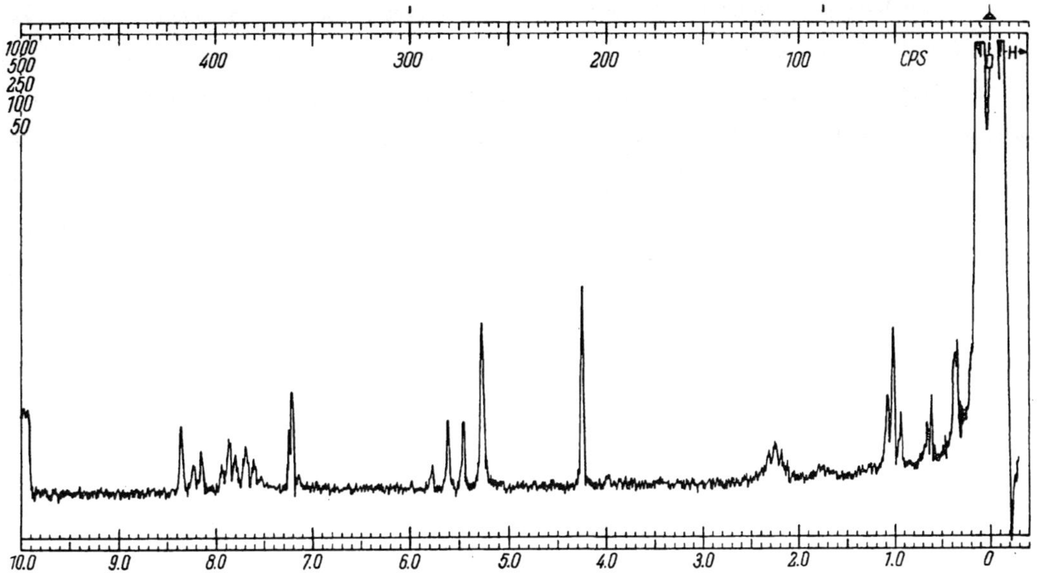

Figure 3. NMR Spectrum of Camptothecin Chloroacetate (100 MG in Deuteriochloroform)

ture VIIIa is in complete agreement with all the physical properties described and has one optically active center. The equivalent form shows the probable relationship of camptothecin to indole alka-

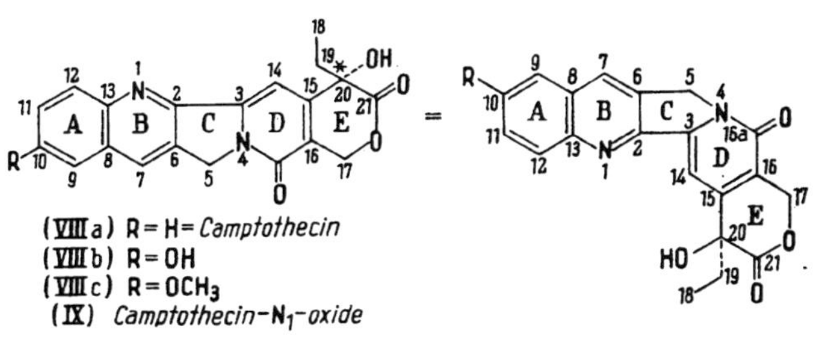

Anti-Tumor Activity

Camptothecin is one of the most active, natural anti-tumor agents. It has very high activity in L-1210 and a variety of other leukemia systems including P-388 and P-1534, the latter being the system in which the Vinca alkaloids were screened. In addition, camptothecin also is very active in inhibition of the growth of solid tumor systems including Walker Carcinosarcoma 256. Camptothecin has survived a variety of preliminary and advanced pharmacological screening in dogs and monkeys. Recently initial trials in man have been initiated.

Structure-Function Relationships

In contrast to the situation with many other anti-tumor agents, we know a great deal about structure-activity relationships in the camptothecin series. Recently hydroxy- and methoxycamptothecin have been isolated from C. acuminata, VIIIb and VIIIc, respectively. These compounds are at least as active as VIIIa. Treatment of VIIIa with metachlorbenzoic acid gives the 1-N-oxide (IX), a compound which is less active than camptothecin but which still has moderate activity. Formation of the sodium salt, Xa or the amide Xb gives compounds with appreciable activity. Xa and Xb are easily converted back to VIIIa at pH 6.0 or less. Changes in the C-20-hydroxyl function lead to complete inactivation. Thus the acetate XIc is essentially inactive, and the C-20-chloro (XIa)- and desoxy (XIb)-analogues are totally inactive. Reduction of VIII under mild conditions with sodium borohydride gives the lactol XII also which is inactive. It is thus apparent that the α-hydroxy lactone in camptothecin is absolutely

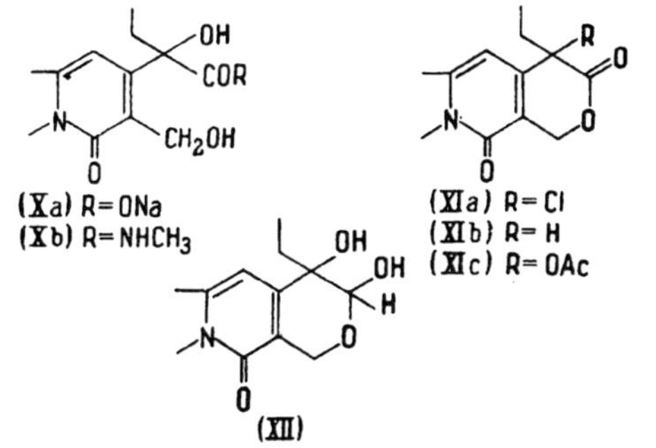

(Xa) R=ONa
(Xb) R=$NHCH_3$
(XIa) R=Cl
(XIb) R=H
(XIc) R=OAc
(XII)

required for activity. Other features of this molecule are under study. It can be rationalized, but remains to be proven that the mode of action involves nucleophilic attack on the R-NH_2 or S-SH groups in enzyme systems or substrates essential to the growth of the various tumors. The α-hydroxy lactone is a more reactive species than a simple lactone. Thus hydrogen bonding of the hydroxyl

group to the lactone carbonyl would increase the positive charge on the carbonyl carbon atom and permit a facile reversible reaction as shown. This also accounts for the fact that VIIIa reacts quantitatively with base, amine, and borohydride under mild conditions.

References

(1) Ebbell, B., 'The Papyrus Ebers. The Greates Egyptian Medical Document'. Oxford University Press, London 1937.

(2) Hartwell, J.L., Lloydia 30, 379 (1967)

(3) Hartwell, J.L., and B.J. Abbott, Antineoplastic Principles in Plants: Recent Developments in the Field, From Advances in Pharmacology and Chemotherapy, Vol. 7, Editor R.J. Schnitzer and A. Golden, 1969, in Preparation, Academic Press

(4) Noble, R.L., Lloydia 27, 280 (1964)

(5) Svoboda, G.H., Lloydia 24, 173 (1961)

(6) Neuss, N., M. Gorman, H.E. Boaz, and N.J. Cone, J. Amer. chem. Soc. 84, 1509 (1962)

(7) Neuss, N., M. Gorman, W. Hargrove, N.J. Cone, K. Biemann, G. Büchi, and R.E. Manning, J. Amer. chem. Soc. 86, 1440 (1964)

(8) Neuss, N., M. Gorman, and N.J. Cone, Lloydia 27, 389 (1964)

(9) Bernard, J., Lloydia 30, 291 (1967)

(10) Frei, E., W.W. Sutow, and J.K. Lace, Lloydia 30, 324 (1967)

(11a) Govindachari, T.R., in: R.H.F. Manske, The Alkaloids, Vol. IX, pp. 517-528. Academic Press, New York 1967

(11b) Govindachari, T.R., Tetrahedron (London) 4, 3111 (1958)

(11c) Govindachari, T.R., M.V. Lakshmikantham, B.R. Pai, and S. Rajappa, Tetrahedron (London) 9, 53 (1960)

(12a) Rao, K.V., R.A. Wilson, and B.M. Cummings (in press)

(12b) Gellerband, F., and R. Rudzats, J. med. Chem. 7, 361 (1964)

(13) Goodwin, S., A.F. Smith, and E.C. Horning, J. Amer. chem. Soc. 81, 1903 (1959)

(14) Gilbert, B., in: R.H.F. Manske, The Alkaloids. Vol. VIII, p. 335-513. Academic Press, New York 1965

(15) Dalton, L.K., S. Demerac, B.C. Elmes, J.W. Loder, J.M. Swan, and T. Teitei, Austral. J. Chem. 20, 2715 (1967)

(16) Curcumelli-Rodostamo, M., and M. Kulka, in: R.H.F. Manske, The Alkaloids. Vol. IX, pp. 133-174. Academic Press, New York 1967

(17) Kupchan, S.M., K.K. Chakravarti, and N. Yokoyama, J. pharmac. Sci. 52, 985 (1963)

(18) Kupchan, S.M., N. Yokoyama, and B.S. Thyagarajan, J. pharmac. Sci. 50, 164 (1961)

(19) Svoboda, G.H., J. pharmac. Sci. 55, 758 (1966)

(20) Svoboda, G.H., Lloydia 29, 206 (1966)

(21) Wall, M.E. et al., AIC-Techrs. Bulletin-367

(22) Wall, M.E., M.C. Wani, C.E. Cook, K.H. Palmer, A.T. McPhail, and G.A. Sim, J. Amer. chem. Soc. 88, 3888 (1966)

(23) Shamma, M., Experientia (Basel) 24, 107 (1968)

Internationales Symposium

Stofftransport und Stoffverteilung in Zellen höherer Pflanzen

Transport and Distribution of Matter in Cells of Higher Plants

Schloß Reinhardsbrunn, 14. bis 19. Oktober 1968
Veranstaltet von der Biologischen Gesellschaft in der DDR, deren Sektion für Physiologie und Biochemie der Pflanzen sowie der Klasse für Medizin der Deutschen Akademie der Wissenschaften zu Berlin und dem Institut für Biochemie der Pflanzen der Deutschen Akademie der Wissenschaften zu Berlin, Halle/Saale

Band a des Symposiumsberichtes: Vorabdruck wissenschaftlicher Beiträge

Herausgegeben von

KURT MOTHES / EBERHARD MÜLLER / AXEL NELLES / DIETER NEUMANN

in Englisch und Deutsch

1968. V, 215 Seiten — 58 Abbildungen — 15 Tabellen — 4° — 24,— M

Das Gebiet des Stofftransportes im Bereich lebender Zellen hat sich in den letzten Jahren zu einem zentralen Kapitel der Molekularbiologie entwickelt. Das Symposium zeigt, in welcher Weise die Untersuchung von Pflanzenzellen dazu beigetragen hat. Die auf dem Symposium behandelte Problematik gruppiert sich in die Abschnitte: Passive Permeabilität, Wasser, Modelle — Verschiedene Aspekte des Transportes — Transport in Chloroplasten und Mitochondrien — Zellpotentiale und Transport — Vitalfärbung und sekundäre Pflanzenstoffe.

Von verschiedenen Standpunkten aus werden Gebiete behandelt, die gegenwärtig einen Höhepunkt der experimentellen Bearbeitung erleben. Daneben wird die Brücke zu den reichen Erfahrungen der klassischen botanischen Zellforschung geschlagen und es werden zukünftige Schwerpunkte angedeutet.

Bestellungen durch eine Buchhandlung erbeten

A K A D E M I E - V E R L A G   B E R L I N

Gibberelline
ihre Derivate und Abbauprodukte

Eine tabellarische Übersicht

Von GERNOT SCHNEIDER / GÜNTHER SEMBDNER / KLAUS SCHREIBER

1966. 232 Seiten – gr. 8° — 19,– M

Seit der Entdeckung der Gibberelline durch japanische Forscher im Jahre 1938 wurden außerordentlich viele Arbeiten über die Physiologie, Biochemie und Chemie dieser Verbindungen veröffentlicht. Besonders interessant — auch im Hinblick auf praktische Anwendungsmöglichkeiten — sind die Gibberelline dadurch, daß sie nicht nur Stoffwechselprodukte des Pilzes Gibberella fujikuroi (Saw.) Woll. sondern auch natürliche Wuchsstoffe der höheren Pflanzen darstellen. Die Vielzahl der im Verlauf der chemischen Erforschung dieses Naturstoffgebietes bis heute bekannt gewordenen, jedoch lediglich zu einem geringen Anteil biologisch geprüften Verbindungen (13 native Gibberelline und über 600 Derivate und Abbauprodukte) ist in dieser Übersicht nach strukturellen Gesichtspunkten geordnet tabellarisch zusammengestellt. Die Tabellen enthalten neben den chemisch-systematischen Bezeichnungen der Verbindungen deren Summen- und Strukturformeln, die in der Literatur angeführten Schmelzpunkte, optischen Drehwerte sowie spektroskopischen Angaben. In einer zusätzlichen Tabelle sind alle Verbindungen nach ihren Schmelzpunkten geordnet.

Bestellungen durch eine Buchhandlung erbeten

A K A D E M I E - V E R L A G B E R L I N

FSC
www.fsc.org
MIX
Papier | Fördert gute Waldnutzung
FSC® C083411

Zeitfracht Medien GmbH
Ferdinand-Jühlke-Straße 7
99095 Erfurt, Deutschland
produktsicherheit@kolibri360.de